Discourses on Sustainability

"Calls for sustainability—including sustainable development and sustainable peacebuilding—have become more forceful in recent years in response to past, present, and anticipated environmental crises, especially those brought about by human-induced climate change. Concomitant have been demands for justice—climate justice, distributional justice, energy justice, environmental justice, procedural justice, social justice, and the protection of human rights. As Dmitry Kurochkin and Elena V. Shabliy assert in the introduction to this powerful volume, human-induced climate change has widened extant social and economic inequality and injustice; global climate change cannot be mitigated without ensuring justice for all."

"The eight chapters of *Discourses on Sustainability: Climate Change, Clean Energy, and Justice* compellingly introduce readers to a variety of narratives on sustainability, climate change, and energy politics from Sweden, Australia, eastern Russia (Yakutia), Uganda, Turkey, and India, as well as from a more global perspective. Also noteworthy about this volume is the diversity of contributors—scholars hailing from Australia, Canada, Finland, India, Sweden, the United Kingdom, the United States, the Virgin Islands, and the public and private sectors. Together they reveal how much has been done but also how much we still need to do to strengthen national efforts and increase international support and global cooperation for mitigating climate change and shattering inequality."

—Karen L. Thornber, *Harry Tuchman Levin Professor in Literature* and *Professor of East Asian Languages and Civilizations, Harvard University*

Elena V. Shabliy · Dmitry Kurochkin ·
Martha J. Crawford
Editors

Discourses on Sustainability

Climate Change, Clean Energy, and Justice

palgrave
macmillan

Editors
Elena V. Shabliy
Harvard University
Cambridge, MA, USA

Dmitry Kurochkin
Harvard University
Cambridge, MA, USA

Martha J. Crawford
Jack Welch College of Business & Technology
Sacred Heart University
Fairfield, CT, USA

ISBN 978-3-030-53120-1 ISBN 978-3-030-53121-8 (eBook)
https://doi.org/10.1007/978-3-030-53121-8

This Palgrave Macmillan imprint is published by the registered company Springer Nature Switzerland AG
The registered company address is: Gewerbestrasse 11, 6330 Cham, Switzerland

Foreword

The adoption of the Paris Agreement's goal to reduce greenhouse gas emissions by every country in the world indicates the extent of attention to climate change. While signing an international accord does not commit a country to concrete action, it is a strong indication of the recognition of climate change as an issue that needs to be addressed and an interest in doing so. A significant step in meeting the emission reduction targets needed to reduce future increases in global temperatures will be to develop a cleaner energy sector. As described by former United States President Obama, the Paris Agreement "sends a powerful signal that the world is firmly committed to a low-carbon future. And that has the potential to unleash investment and innovation in clean energy at a scale we have never seen before."[1]

Scaling up clean energy can contribute to climate change goals, directly to the mitigation objective of reducing emissions so as to control future increases in temperatures, and indirectly to adaptation and resiliency measures by reducing the impacts of climate changes. Looking beyond climate change, there are also economic opportunities as new industries and work forces develop to fulfill the demands of a clean energy sector. How programs that mandate or support movement away from fossil fuels are designed will affect the extent to which new opportunities can offset negative impacts, such as job losses, that come from this shift.

At the same time, there is a need to ensure that the transition away from fossil fuels and development of the clean energy sector is equitable. Adding this layer of analysis can be difficult in an already complicated decision-making process. As Judith Herbst and Deanna Grant-Smith note in Chapter 3, responding to climate change and ensuring energy security are often framed as competing policy interests. As the essays in this book illustrate, advancing these goals simultaneously requires diverse partnerships and integrating equitable concerns may implicate additional challenges and opportunities.

In this context, several of the book's authors discuss how equity can be measured from many perspectives; three that are commonly discussed in the climate change context are environmental justice, energy justice, and climate change justice. These have commonalities in their goals, such as looking to protect members of society that may (i) be most at risk, either from harm from pollution, lack of access to energy, or negative impacts of climate change, but (ii) often have less contribution, on average, to the source of the risk. Similarly, each approach to equity may try to redress both historic and ongoing disparities in distributions of risks and access to benefits.

While there are similarities between these forms of equity, there can also be differences, ranging from the people they seek to protect, the harms they seek to avoid, and the strategies they employ to achieve fair results. For example, environmental justice policies typically seek to protect or assist people at the community level, defining protected neighborhoods by reference to factors such as income levels and percentages of minorities, foreign-born residents, and/or non-English speakers. However, this approach may not be sufficient to protect individuals in their role as energy consumers as opposed to in their role as community members. Wealthy communities can have residents with income levels below designated poverty levels who would need assistance with their energy bills; this is particularly true in countries that mandate or incentive broad dispersion of affordable housing. Even when proponents of environmental protection and clean energy seek to protect the same people, they may disagree on what that means. To promote equitable solutions and innovations, advocates, and decision-makers should consider both environmental and energy equity perspectives in order to identify and take advantage of synergies and resolve potential conflicts. Because energy justice is a newer concept, its framework is discussed briefly here.

Energy justice can be measured at many levels. At perhaps the highest level, a fundamental question of equity is whether everyone has access to the basic level of energy that modern society deems necessary for purposes of safety, health, and economic growth. As impacts of climate change lead to greater fluctuations in temperature this baseline need for energy may increase to meet growing needs for air conditioning and heating. This step in the equity analysis is particularly relevant when considering energy justice from a global perspective. According to the World Bank, approximately 11% of the world's population did not have access to electricity in 2017, with a more significant lack in rural than urban areas.[2] From a purely economic standpoint, the historical application of this principle would have frequently focused on access to the lowest-cost forms of energy. When starting from a concern about climate change, this analysis may look beyond direct costs to account for environmental implications, including long-term ability to adapt to climate change impacts. While this could increase benefits, it could also lead to higher costs, even if only temporarily, for some or all energy consumers.

The next stage of an energy justice analysis may consider the direct cost of access to energy, *i.e.,* the purchase price. A proxy for this issue is a consumer's, or household's, energy burden. From an economic perspective, energy burdens refer to the percentage of income households spend on energy costs. Studies in the United States indicate that low-income households generally have higher energy burdens than other households, primarily because their income is lower but also because their homes tend to be older and less energy-efficient.[3] Energy burdens are problematic not just because of their absolute costs but because they can also force households to make tradeoffs between energy and other basic provisions, like food, medicine, and rent, that can jeopardize health, safety, and housing stability. Energy burdens could be particularly significant as the world works to close energy gaps in less developed regions; high infrastructure costs to reach remote areas would, unless subsidized by a government, be passed on to consumers in the price of their energy. This is just one reason that looking beyond traditional systems is important as we think how to build a clean energy sector that can support climate change goals while also promoting equity.

The third phase of an energy justice analysis may evaluate the negative impacts and benefits, and potential co-benefits, associated with the energy available to different individuals. This entails considering the type of energy to which people have access, *i.e.*, whether the underlying fuel is coal, oil, or some other fossil fuel versus a renewable source such as solar, wind, or hydro. In many ways, not all energy is created equal. Fuel type can have impacts, both negative and positive, related to issues such as reliability, public health, the environment, and local economies. For example, a neighborhood powered solely by solar panels may have less air pollution than a community powered by a coal plant, but may have more interruptions in power due to the intermittent nature of renewables (unless paired with storage technologies). Evaluation of impacts should look beyond fuel choice and generation and also consider the impacts of supporting infrastructure, such as transmission and distribution systems. For instance, the public health and environmental impacts of natural gas pipelines and wind turbines are different, and solar power via individual rooftop systems versus community "gardens" present different opportunities for consumer participation.

A final element of an energy justice analysis may consider the comparative distribution of the negative aspects of energy choices and access to benefits associated with the energy system. This differs from the third step's consideration of *absolute* impacts of the energy sector on individuals. Such benefits could take the form of reduced environmental impacts, greater control over electricity usage, or reduced bills. Applying this principle may require considering whether costs and benefits should be distributed equally to all consumers, proportionally based on contributions to cost, or disproportionately favoring those consumers most unable to afford costs or most "in need" of the benefits. When discretionary actions result in benefits, systems should be designed to enable as many people as possible to take the beneficial actions.

Putting these together, the substantive goals of energy justice could include:

1. Assuring access to reliable sources of energy;
2. Avoiding and reducing energy burdens, particularly disparate burdens, on low-income consumers;
3. Minimizing the imposition and disproportionate distribution of the costs or negative impacts associated with building, operating, and maintaining energy generation, transmission, and distribution systems; and
4. Providing equitable distribution of and access to real benefits associated with building, operating, and maintaining energy generation, transmission, and distribution systems.

While all aspects of this framework are worthy objectives, simultaneously making progress on all principles will not always be possible, but acknowledging what we can and cannot accomplish is still an important step.

This energy justice framework focuses on substantive/distributive justice as opposed to procedural/participatory rights. But as several of the essays in this book illustrate, implementing a clean energy system that integrates equitable goals requires input from many sectors, including legislative bodies and agencies at all levels of government, courts, the public, advocacy groups, and the private sector. This need for coordination and broad support underscores the value in addressing equity with respect to participation in relevant decision-making. Ability to participate requires both opportunities to do so and capacity. Developing a clean energy sector involves complex issues; improving energy literacy can support more informed participation in decision-making by all participants, including advocacy groups and individuals.

The move to a cleaner energy sector is an important step in addressing climate change that raises fundamental questions about how to balance innovation with costs to individuals. If approached with a proactive goal of addressing issues of equity, this transition presents opportunities to create systems that learn from prior practice and iteratively improve the equitable impacts of the energy sector.

Aladdine Joroff
Emmett Environmental Law and Policy Clinic
Harvard Law School
Cambridge, USA

Notes

1. Statement by the President on the Paris Climate Agreement (Dec. 12, 2015) https://obamawhitehouse.archives.gov/the-press-office/2015/12/12/statement-president-paris-climate-agreement.
2. World Bank, Access to Electricity (% of population), https://data.worldbank.org/indicator/EG.ELC.ACCS.ZS.
3. Rental households also experience higher energy burdens. See, e.g., Ariel Drehobl & Lauren Ross, American Council for an Energy-Efficient Economy, Lifting the High Energy Burden in America's Largest Cities: How Energy Efficiency Can Improve Low Income and Underserved Communities 12 (2016).

Aladdine Joroff practices environmental law at Harvard Law School's Emmett Environmental Law and Policy Clinic with a focus on climate change mitigation and adaptation strategies for municipalities and environmental and energy justice. She also teaches environmental law and policy courses at several of Harvard University's schools.

Contents

Editors and Contributors

About the Editors

Elena V. Shabliy graduated with honors from M.V. Lomonosov Moscow State University and received her Interdisciplinary Ph.D. from Tulane University in 2016. In 2009, she earned a Master of Liberal Arts degree from Tulane. She was a Visiting Scholar at Harvard University in 2015–2017 and Columbia University in 2017–2019. She is the editor of *Representations of the Blessed Virgin Mary in World Literature and Art* (Lexington, Rowman and Littlefield, 2017) and co-editor of *Emancipation Women's Writing at Fin de Siècle* (Routledge, 2018) and *Renewable Energy: International Perspectives* (Palgrave Macmillan, 2019). She was a Postdoctoral Fellow at Harvard University in 2018. Currently, she is a Visiting Scholar at the Davis Center for Russian & Eurasian Studies at Harvard University.

Dmitry Kurochkin is a Lecturer and Senior Research Analyst at Harvard University. He graduated *magna cum laude* from Lomonosov Moscow State University where he was majoring in Physics. Kurochkin earned his Ph.D. in Mathematics at Tulane University and holds Master's degrees in Chemistry, Statistics, and Economic Analysis and Policy.

Martha J. Crawford, Ph.D. is Dean of the Jack Welch College of Business & Technology at Sacred Heart University. She has both business and academic experience. She has served as senior vice president of Research & Development for several companies based in France such as L'Oreal and Air Liquide, where she has worked to foster technological innovation. Her academic background includes having taught a MBA core-curriculum course at the Harvard Business School on leadership and corporate responsibility in addition to developing a course on energy that was nominated for a national prize. Crawford earned her master's degree in business administration from the Collège des Ingénieurs in Paris and MS and doctorate degrees in environmental and chemical engineering from Harvard University. She has also served as a board member in both academic and professional settings for Sevres International School, French National Labs for Scientific Research and French National Energy Labs and board director for Altran, Suez and Ipsen all in France. Crawford also has several publications focusing on environmental management and quality.

Contributors

Dr. Shilpa Daithota Bhat is assistant professor in Ahmedabad University, Gujarat, India. She is the recipient of the *Bensley-Osler Library Research Travel Grant* (McGill University, Canada, 2017); and the inaugural *Lorna Marsden International Visitor Fellow* (York University, Canada, 2015). She was a *Commonwealth Fellow* (University of Toronto); and recipient of the *Pacific Asia Network of Canadian Studies* (Korea University, Seoul, South Korea). Her areas of interest are Indian Diaspora, South Asian Narratives, South Asian Canadian Literature, Postcolonial theories, and Canadian Studies. She is the author of *Indians in Victorian Children's Narratives: Animalizing the 'Native', 1830–1930* (Rowman and Littlefield, Lexington Books, US 2017). Her edited anthology is entitled *Diaspora Poetics and Homing in South Asian Women's Writing: Beyond*

Trishanku, (Rowman and Littlefield, Lexington Books, US, 2018). She is the co-editor of *Women Writers of the South Asian Diaspora: Interpreting Gender, Texts, and Contexts* (forthcoming with Rawat Publishers, Jaipur).

Nisha Clavier, MAE is an Instructor of Social Sciences at the University of the Virgin Islands.

Kimarie Engerman, Ph.D. is a full Professor of Psychology, and Dean of the College of Liberal Arts and Social Sciences at the University of the Virgin Islands.

Jessica García-Terán is a Ph.D. student at Department of Engineering Sciences, Industrial Engineering & Management, Uppsala University. Her main study is devoted to how the Swedish discourse on sustainability affects local governing and regional development in the knowledge economy.

Deanna Grant-Smith has more than 20 years of experience conducting research and providing policy and strategy advice across the areas of environmental policy, participatory planning, community engagement, and organizational strategy and development. She has extensive experience in developing and delivering consultative approaches for a range of issues including transport and infrastructure planning, and environmental policy review. Grant-Smith has conducted and supervised a range of research activities to inform strategic evidence-based decision-making. Currently, she researches on public involvement and stakeholder engagement with a focus on marginalized or disadvantaged groups, particularly young people and women; and she investigates cultural and political engagements linked to environment and sustainability issues, particularly those associated with waste.

Dr. Brett Grimm completed a Ph.D. at The University of Adelaide 2009. The topic of Brett's thesis is "Quantifying the Visual Effect of Wind Farms; A Theoretical Process in an Evolving Australian Visual Landscape." Since 2003, Brett has worked in the private and public sector in Australia and the United Kingdom as a Landscape Architect, Urban Planner, and Strategic Project Senior Advisor. Brett's research interests in renewable energy landscapes and landscape visual amenity have provided a focus for private consultancy and the development of the Grimke matrix. The Grimke matrix (employed and co-authored by BGLA and WAX Design) is a visual assessment methodology that has been used on numerous proposed wind farms, solar farms, mine expansions, desalination plants, and associated road corridor infrastructure projects within Australia. The Grimke matrix was awarded an Australian Institute of Landscape Architect (SA) Infrastructure award of excellence (2016). Brett has provided expert advice on numerous visual assessment cases within the Environmental Resource and Development Courts (SA). Brett is an active participant in the Australian Institute of Landscape Architects education committee, being the past chair of South Australia and a member of a recent national accreditation review team.

Daria Gritsenko is Assistant Professor at the University of Helsinki, Finland, affiliated with the Aleksanteri Institute and the Helsinki Center for Digital Humanities (HELDIG). Her scholarly focus is on policy-making activities of private actors, sustainability, and governance in the digital world. Her work has been published in major international peer-reviewed journals, including *Energy Policy*, *Transport Policy*, *Geoforum*, *Policy Studies Journal*, and professional magazines. She is co-founder of Digital Russia Studies, an interdisciplinary network of scholars working at the intersection of "digital" and "social" in Russia and beyond.

Judith Herbst lectures this semester on Business in Australia and Business in Europe in the School of Management at QUT Business School, and she conducts research on Managing Sustainable Change. Her research interests are aligned with how organizations integrate change using systems to co-create value with stakeholders along the value chain. Previously, Herbst worked in industry as a marketing communications specialist where she completed projects for commercial and non-profit organizations, including the College Community Redevelopment Project in San Diego and Green Cross Australia.

Sharon Honore, Ph.D. is an Associate Professor of Communications at the Associate Professor of Communications.

Çiğdem Pekar has a B.A. degree from the Department of International Relations at Ege University, Turkey, an M.A. degree on "European Studies" from the University of Exeter, UK and a Ph.D. degree in International Relations from Çanakkale Onsekiz Mart University, Turkey. She has spent one academic year at the Center for NonProliferation Studies, Monterey, US as a Fulbright Ph.D. researcher. Currently she is pursuing her career at Çanakkale Onsekiz Mart University as an Assistant Profesor. Dr. Pekar's research areas include nuclear nonproliferation regime, Turkey's nuclear and renewable energy policies, and nuclear history. She is a member of several academic and professional societies and organizations such as: International Young/Student Pugwash Group, Women in International Security (WIIS), Women in Nuclear (WIN), and International Nuclear Law Association (INLA). Dr. Pekar also serves as the Institution Representative for the IAEA International Nuclear Security Education Network (INSEN) on behalf of her University.

Annika Skoglund is Associate Professor at Uppsala University, Sweden and The University of Exeter, UK. Her main research area is new organizational forms, with focus on how these arise and affect businesses, technological paths, and the human. She has published broadly on sustainability and climate change, and specifically on renewable energy activism within corporations and smaller businesses. She is also developing videographic methods for the study of "alternative entrepreneurship," such as social, green, political, and sustainable entrepreneurship.

Małgorzata Zachara, Ph.D. is an associate professor at the Faculty of International and Political Studies, Jagiellonian University in Poland, working on social change, environmental sustainability, governance of global processes, and political leadership, with emphasis on individual empowerment. She has accumulated several years of management and academic experience lecturing in different European universities and coordinating several scientific and business-oriented projects. Her recent publication includes: *The Millennial Generation in the Context of Political Power: A Leadership Gap?*, "Leadership" 2020; *Can Institutions Save the World? Neoliberal Institutionalist Perspective on Global Governance*, "Stosunki Międzynarodowe/International Relations," vol. 55 nr 1; *Mechanisms of Transformation in the Global Sphere. Centre of Gravity (CoG) as an*

Instrument of Analysis of Global Decision-Making, Jagiellonian University Press 2015; *Global Governance*, (Jagiellonian University Press, 2012).

Joshua Zeunert is a Senior Lecturer at the University of New South Wales in Sydney and a Registered Landscape Architect (AILA). He has worked at five universities internationally, as well as in industry with leading Australian landscape architecture and urban design practices on multiple award-winning projects. His research interests explore on the nexus of academia and professional practice in environmental design strategy, regenerative landscapes, and food systems. He is the author of three books, which have all received professional institute awards, including *Landscape Architecture and Environmental Sustainability* for Bloomsbury Publishing (2017) and the *Routledge Handbook of Landscape and Food* (with Tim Waterman, 2018).

List of Figures

List of Tables

1

Introduction: Climate Ethics and International Energy Justice

Dmitry Kurochkin and Elena V. Shabliy

In the recent past, multiple discourses on sustainability appeared, shedding light on urgent problems of human-caused climate change and clean energy advancements. The concepts of *sustainability* and *sustainable development* have become popular among local communities, international policymakers, and researchers. In addition to these important topics, such themes as climate justice, environmental justice, global energy justice, ecological justice, sustainable justice, and procedural justice remain attractive to scholars and researchers internationally (Joroff 2017; Sovacool and Dworkin 2014; Walker 2011). Sustainable development has become an influential discourse worldwide. In this edited volume, scholars elaborate on various responses to human-induced climate change, calling for action, and encouraging further thorough analysis and research. Human activities have been the dominant cause

D. Kurochkin (✉) · E. V. Shabliy
Harvard University, Cambridge, MA, USA
e-mail: dkurochkin@fas.harvard.edu

E. V. Shabliy
e-mail: eshabliy@tulane.edu

E. V. Shabliy et al. (eds.), *Discourses on Sustainability*,
https://doi.org/10.1007/978-3-030-53121-8_1

of climate warming.[1] At the national and local levels, there are various reactions to these problems, such as, for example, the National Climate Assessment (NCA), the Voluntary National Reviews (Chapter 6), the Mandatory Renewable Energy Target, the Large-scale Renewable Energy Target (Chapter 5).[2] At the macro-level, the recognition of these existing problems was supported by the United Nations, the UN Earth Summit, the Rio+20 Summit, the Kyoto Protocol, the Paris Agreement, the Intergovernmental Panel on Climate Change (IPCC), the Bali Principles of Climate Justice, etc.[3] Nationally Determined Contributions (NDCs) are at the core of the Paris Agreement and the achievement of numerous goals.[4] NDCs represent those efforts by each country and region to cut national emissions and adapt to climate change.[5] The 2030 Agenda for Sustainable Development was introduced in 2015; it has 17 Sustainable Development Goals (SDGs) and 169 associated targets.[6] The world is working toward all these Goals, including Goal 7 (Affordable and Clean Energy), Goal 6 (Clean Water and Sanitation), Goal 13 (Climate Action), Goal 16 (Peace, Justice, and Strong Institutions), and Goal 17 (Partnership for the Goals), which are considered in this edited volume.

This interdisciplinary book presents a critical reflection on climate change (Chapters 1, 2, 3, 4, 6, and 7) and clean energy (Chapters 5, 6, 8, and 9), which inevitably draw on fundamental philosophical inquiries of social justice and human rights problems. Climate change is expected to negatively impact ecosystems in the future; "a public action approach to policy provides a better lens than the conventional rationalist approach to analyse the contested nature of climate science and the potential of lived experience to inform debates through active engagement."[7] Diverse knowledge of and approaches to climate change help understand this growing problem. Global average air temperature has increased in the recent past by approximately 1.0 °C (1.8 °F); according to the Climate Science Special Report, the last several years have been record-breaking, and the period of 1901–2016 is the warmest.[8] Greenhouse gas (GHG) emissions are still rising, with damaging effects on the Earth's climate.[9] At the moment, the concentration of CO_2 is higher than at any point in time—at least the past 800,000 years.[10] However, carbon dioxide (CO_2) is not the only GHG that impacts human-induced climate change:

"Concentrations of methane (CH_4) and nitrous oxide (N_2O), even more potent greenhouse agents that CO_2, are also at record high level and likely to increase in the future."[11]

Abbott and Wilson distinguish between *lived experiential* and *local/indigenous* knowledge and the role of knowledge in policy-making on climate-related problems (Abbott and Wilson 2014). *Lived experience* is a process of evolution, often over many years of generations.[12] Abbott and Wilson argue that there are countless unique *lived experiences* when it comes to climate change problems.[13] Moreover, *lived experience* is characterized by its diversity, and with diversity sometimes comes disagreement, then dialogue, and possible solution.[14] The role of public participation and the development of sustainable communities are two very important aspects; local sustainability projects are still top-down.[15]

Climate change is evident in many forms, such as, for example, the most obvious—recent weather fluctuations that happen around the world. Floods, droughts, and hurricanes are those visible signs of climate change. Human-caused climate change is projected to greatly impact marine, freshwater, and terrestrial life.[16] Damage estimates after hurricane Sandy in 2012 range as high as $60 billion.[17] Rising sea levels and its negative impact is widely discussed; global average sea level has risen by about 7–8 inches since 1900.[18] Temperatures in Alaska and the Arctic have increased over the last 50 years "at a rate more than twice as fast as the global average temperature."[19] The Arctic Ocean is losing the summer ice cover: "prospects are that this region will be largely ice-free for at least part of the year within a few decades."[20] Greenland ice sheets are losing its thickness as well.[21] Poor people are more vulnerable to man-made climate change and respond rapidly to its impacts. They have no choice because their livelihoods and well-being depend on such responsive ability. Sometimes their response results in other problems; for example, whole rural communities abandon their present locations and migrate to the nearest places, generating new complications, as has been recorded in coastal areas of Bangladesh after flooding.[22] Many people, however, adapt to their changed circumstances. Examples in poor rural areas include selectively breeding seeds with climate variability resistance and building houses on stilts on the Pacific Islands to protect against floods.[23]

The term sustainability has a broad meaning, including sustainable *peacebuilding*; a growing number of researchers in the field proves that the questions of sustainability and renewable energy are very important and leave no space for social indifference and inertia. The idea of regenerating power itself is very natural and is expected to remain attractive until its full realization and implementation worldwide.[24] At the end of the nineteenth century, Anton Pavlovich Chekhov writes about the spiritual roots of the environmental crisis in his play *Uncle Vanya* (1898). Chekhov's dream was to become a gardener, and his care for the environment and nature is evident throughout his work. Writers, scientists, and spiritual leaders are deeply concerned about the current environmental changes. Climate change is not just an environmental issue; it is an urgent problem and ethical issue. This global phenomenon requires international cooperation and common efforts by people of many cultures and religions living together.[25] Consequences include but not limited to social turbulence and economic disbalance. It should be mentioned that *all* social strata will experience the consequences of human-induced climate change. Governments reached a universal climate change agreement in Paris. According to this agreement, the average global temperature should be kept from rising beyond 2 °C. At the UN General Assembly, in New York, there was a gathering to reach a new set of sustainable goals.

The Catholic Church views climate change as an ethical issue that must be addressed in order to protect the planet.[26] Pope Francis participates in the fight against climate change and also called everyone to join this environmental battle. His Holiness Pope Francis released the papal encyclical. This address entitled *LAUDATO SI'* underlines that climate change is one of the main current and future challenges.[27] In this letter, Pope Francis reminds that, according to Saint Francis of Assisi, our common home is like a "sister with whom we share our life and a beautiful mother who opens her arms to embrace us."[28] He writes that the sister now cries out because of all the harm we have inflicted on the planet by our irresponsible use and abuse of the goods reminding that our bodies are made up of the earth's elements.[29]

Countries around the world participate in climate change meaningful dialogue. Sweden was successful in breaking the link between economic growth and GHGs (Chapter 2).[30] The Swedish government adopted sixteen environmental objectives as well as the objective of reduced climate impact.[31] The reduction of GHGs is important in human-caused climate change mitigation action plan:

> Early greenhouse gas emissions mitigation can reduce climate impacts in the nearer term (such as reducing the loss of arctic sea ice and the effects on species that use it) and in the longer term by avoiding critical thresholds (such as marine ice sheet instability and the resulting consequences for global sea level and coastal development; Ch. 29: Mitigation, Timing and Magnitude of Action).[32]

In 2008, in Swedish town Uppsala, the Archbishop of the Church of Sweden, Anders Wejryd, invited 30 religious leaders to attend an Interfaith Summit on Climate Change. As a result, there was issued an Interfaith Climate Manifesto. Countries must, as the Manifesto states, ensure that vulnerable communities are supported. In this Manifesto, there is a call for effective leadership and action in view of the global threat: "Ethics and values are intrinsic to the development of new institutional structures and architectures of politics and finance. In the religious realm long-sightedness has always been important. More than ever before the world now needs extraordinary, long-sighted political leadership."[33] In the Interfaith Climate Manifesto, there is an encouragement for global political leadership. Developed countries, especially those in Europe and North America, must lead the way in cutting emissions; in the developed countries emissions should be reduced by at least 40% by 2020 and 90% against 1990 levels.[34] The climate change problem is linked to "excessive consumption of the wealthy lifestyle."[35] Both—developed and developing countries—are responsible for the historical GHG emissions but the wealthiest countries "generally have the greatest per capita emissions."[36] The Manifesto highlights an urgent need for "economic incentives for developing countries to foster cleaner development on a national scale."[37] This Summit also recognized the goal that was articulated in Paris—to keep from rising beyond 2 °C the average global temperature. On September 1, 2017, Pope Francis and Patriarch Bartholomew issued a joint message on the World Day of Prayer for Creation: "We no longer respect nature as a shared gift; instead, we regard it as a private possession. We no longer associate with nature in order to sustain it; instead, we lord over it to support our own construct. The human environment and the natural environment are deteriorating together, and this deterioration of the planet weighs upon the most vulnerable of its people."[38] Pope Francis and Patriarch Bartholomew conclude their message, appealing to those in positions of social and economic as well as political and cultural responsibility.

The Russian Orthodox Church, the Patriarch Kirill of Moscow supports the climate change dialogue. Russia joined the Paris Agreement on Climate Change and "formally ratified the Paris agreement in October 2019."[39] Renewable energy investments have recently increased in Russia.[40] His Holiness the 14th Dalai Lama participates in this important and timely discussion. The Jewish spiritual community signed a declaration of reducing GHGs; it was supported by the Coalition on the Environment and Jewish Life. Israeli Orthodox Rabbi Yonatan Neril, founder and director of the Interfaith Center for Sustainable Development, is an interreligious environmental advocate. Islamic spiritual leaders play an active role in climate change dialogue.[41] In the Hindu tradition, there is a short Sanskrit prayer that some Hindus recite acknowledging Mother Earth.[42]

Protecting Human Rights in the Context of Climate Change

The Stern Review refers to climate change as an example of market failure, and climate change is expected to have a serious impact on the enjoyment of human rights worldwide; the interrelation of human rights and human-induced climate change is patent—according to the IPCC, "predicted changes in climate will adversely affect billons of people and the ecosystems, natural resources, and physical infrastructure upon which they depend."[43] In 2015, "Climate Change and

Human Rights" Columbia University Law School report describes negative climate-related impacts that will affect individuals and communities. It is important to protect individuals from climate change harm and cooperate internationally to protect human rights against these impacts:

- The impacts of climate change on freshwater resources, ecosystems, and human settlements are already undermining access to clean water, food, shelter, and other basic human needs; interfering with livelihoods.
- These impacts constitute a serious interference with the exercise of fundamental human rights, such as the rights to life, health, water, food, housing, and an adequate standard of living.
- Mitigation, adaptation, and geoengineering measures can also adversely affect the exercise of human rights.[44]

Energy justice has recently emerged as social science agenda, and it seeks to apply justice principles to energy policies, energy consumption, energy production and systems, energy activism, and energy security.[45] Energy justice evaluates situations when injustices may emerge; it begins with questioning "the ways in which benefits and ills are distributed."[46] Aladdine Joroff writes: "Energy justice is a relatively new concept as compared to environmental justice, and although the ideas are related, they at times diverge in objectives and strategies."[47] Michael H. Dworkin and Benjamin K. Sovacool in *Global Energy Justice: Problems, Principles, and Practices* give the following definition of energy justice:

> Energy justice, thus, involves the right of all to access energy services, regardless of whether they are citizens of more or less greatly developed economies. It encompasses how negative environmental and social impacts related to energy are distributed across space and time, including human rights abuses and the access disenfranchised communities do or should have to remedies. Energy justice ensures that energy permitting and siting do not infringe on basic civil liberties and that communities are meaningfully informed and represented in energy decisions.[48]

John Rawls defines justice in *A Theory of Justice* (1971) as "the first virtue of social institutions, as truth is of systems of thought. A theory however elegant and economical must be rejected or revised if it is untrue; likewise laws and institutions no matter how efficient and well-arranged must be reformed or abolished if they are unjust."[49] The question of energy distribution is of great importance; in Germany, for example, the renewable energy transition (or the *Energiewende*) has occurred after the passage of the Renewable Act in 2000 by the German Federal Government.[50] The *Energiewende* comes with distributional justice implications and entails the decarbonization of the German energy sector, as well as the removal of nuclear power from the energy mix.[51] The German Federal Government established Feed-in-Tariffs (FITs), thus enlarging the "number of players in the energy market by giving priority for both *grid access* and *market access* to generators of renewable energy over those producing fossil fuel-based energy."[52]

This edited volume brings together researchers from various regions—Africa, Australia, India, Sweden, Russia, Turkey, and the United States—to discuss the existing and potential problems of climate change and active mitigation and adaptation approaches. Sustainability has been taken very seriously for a long time in the Swedish state and in Scandinavia. Chapter 2 discusses the Swedish government attempts to create citizen awareness of human-caused climate change. Chapter 3 concentrates on climate justice and the pursuit of it through litigation. The impacts of climate change and unmet energy security are inextricably linked, as Judith Herbst and Deanna Grant-Smith argue. This chapter considers legal challenges at multi-levels; it looks at the related and growing cluster of court cases against fossil fuel companies and governments for their failure to protect people globally. 2017 marked the tenth anniversary of *Massachusetts v. EPA* that is a US Supreme Court Case and "one of the most important environmental cases" in the USA.[53] In 2006, Massachusetts and other eleven states petitioned the Environmental Protection Agency (EPA), asking EPA to regulate GHG emissions from new motor vehicles that cause the global warming and climate change.[54] Massachusetts asked EPA to regulate GHGs by the Clean Air Act. EPA denied the petition reasoning that:

(1) the Act does not authorize it to issue mandatory regulations to address global climate change, and (2) even if it had the authority to set greenhouse gas emission standards, it would have been unwise to do so at that time because a causal link between greenhouse gases and the increase in global surface air temperatures was not unequivocally established.[55]

Later Massachusetts appealed the denial of this petition.[56] The Supreme Court stated that "protection of human health and the environment from air pollution under nation's clean air laws—including protecting the millions of Americans afflicted by the clear and present danger of climate change—must be rooted in science, not politics or expediency."[57] The Supreme Court found that a "well-documented rise in global temperatures has coincided with significant increase in the concentration of carbon dioxide in the atmosphere."[58] Climate change denial and inertia violates rights of young people. *Juliana v. United States* (2016, Nov. 10. No. 6:15-cv-01517, (D.Or.), 46 ELR 20175) is another climate-related lawsuit that was filed in 2015 and dismissed after a 5-year process in 2020. James Hansen writes: "We must expose the facts [of climate change] rigorously so that the courts can protect the rights and future of young people."[59] Procedural justice manifests itself as a "call for equitable procedures that engage all stakeholders in a non-discriminatory way."[60] Procedural justice is more than simply inclusion: "It involves also the mobilization of local knowledge."[61] Procedural justice is concerned with how exactly decisions are made in the determination of social goals, "or who is involved and has influence in decision-making."[62] According to Michael H. Dworkin and Benjamin K. Sovacool, it has four elements: (1) access to information; (2) access to and meaningful participation; (3) lack of bias on the part of decision-makers; and (4) access to legal process for achieving redress.[63]

In Chapter 4, Kimarie Engerman, Nisha Clavier, and Sharon Honore examine the relationship that existed between the awareness of climate change and environmental stewardship of local residents of St. Croix, Virgin Islands. Electronic and paper survey were administered to 317 participants at two local beaches; results from the survey indicated that there were significant associations between certain environmental stewardship behaviors and participants' knowledge and awareness of climate change and global warming. According to the NCA Report, a new National Global Change Research Plan (2012–2021) includes elements related to climate adaptation, such as "improving basic science, informing decisions, improving assessments, and communicating with and educating the public."[64]

In Chapter 5, Joshua Zeunert and Brett Grimm present a case study of wind and solar energy landscapes in the state of South Australia (SA) focusing on visual and strategic planning concerns. Renewable energy sources are those recent means of adaptation and mitigation to climate change that will gradually replace fossil fuels. Solar power is expected to replace fossil fuels like, for example, coal in the near future, according to a Bloomberg New Energy Finance outlook.[65] In the US, Germany, India, and China rapid solar power development already competes with coal. According to this report, through 2040, China and India represent the "biggest markets for new power generation," drawing about 39% of "all investment in the industry."[66] Moreover, cost of electricity generated from photovoltaic panels is expected to fall significantly by 2040. Solar costs will fall in the United States by 67% and by 85% in Japan.[67] The data demonstrate that, despite the common perception that alternative energy use is expensive in comparison with fossil fuels, solar and wind energy are becoming major sources of electricity.[68] The report states: "By 2040, wind and solar will make up almost half of the world's installed generation capacity, up from just 12 percent now, and account for 34 percent of all the power generated, compared with 5 percent at the moment."[69]

Chapter 6 presents sustainability discourses of the Turkish government. Çiğdem Pekar argues that Turkey has been highly dependent on foreign energy resources to meet its energy needs for a long time. As this chapter demonstrates, Turkey has a very strong commitment to renewable energy development and climate change mitigation and adaptation. According to the official documents, Pekar adds, Turkey targets a 30% share of renewable energy and 10% share of nuclear energy in the electricity mix by the date of 2023. In Chapter 7, Małgorzata Zachara considers climate change as an economic issue as well. In Chapter 8, Daria Gritsenko discusses sustainable energy programs and policies in Yakutia, Russia. Since 2013, the Russian federal government adopted new policy instruments in support of renewables, aiming to improve domestic energy security and, according to the official strategies, mitigate climate change. At the Skolkovo Innovation Center, in 2018 there was IV International Congress "Renewable Energy—XXI" in Moscow. Participants discussed the global energy transition to green technologies and the consequences of this transition for economy, energy, and ecology.[70]

Finally, in Chapter 9, Shilpa Daithora Bhat analyzes Manubhai Madhvani's autobiographical work *Tide of Fortune* and representation of renewable energy in this reminiscent writing that is a literary contribution to the discourse of sustainability.

Individual States and regions in the United States aim at clean energy transition. Thus, for example, Hawaii, California, Maine, Massachusetts, Nevada, New Jersey, Virginia, and Washington continue to prioritize shifting to cleaner energy sources.[71] In the last year, the number of solar jobs in Virginia has increased—by approximately 10%—to 3890 jobs.[72] Virginia has an ambitious plan to produce 30% of Virginia's electricity from renewable energy sources by 2030; and 100%—by 2050.[73] By 2028, Virginia has a goal of achieving 5500 MW of wind and solar energy.[74] Commonwealth of Virginia Office of the Governor Executive Order Number Forty-Three (2019) highlights the importance of Energy Equity: "The plan shall also address issues related to equity and *environmental justice* so that the clean energy and climate goals outlined in this Order are achieved in a just manner that advances social, energy, and environmental equity. These clean energy resources shall be deployed to maximize the economic and environmental benefit to underserved communities while mitigating any impacts to those communities. The Plan shall include measures that provide communities of color and low- and moderate-income communities access to clean energy and a reduction in their energy burdens."[75] The Virginia Clean Economy Act will "create the Commonwealth's first Clean Energy Standard…to transition…[the] electric grid to 100% clean energy by 2050 into Virginia law."[76]

The environmental group Sierra Club informs that across the United States over 150 cities, more than 10 counties and seven states, have adopted 100% renewable energy commitment.[77] Six cities and in the United States—Aspen, Burlington, Georgetown, Greensburg, Rockport, and Kodiak Island—already generate 100% of the energy "used community-wide from clean, non-polluting and renewable sources."[78] Many US cities combat climate change through such initiative as the Compact of Mayors, We Are Still In, and by establishing their own Climate Action Plans.[79] The "Ready for 100" Campaign in the United States supports community commitment and dedication to clean energy transition.[80]

Conclusion

Sustainable development has become an influential discourse worldwide.[81] The world of energy inevitably involves ethical issues.[82] Kirsten Jenkins et al. argue that the inclusion of knowledge, discourse and stories in our decisions can make an impact on policies.[83] Climate is changing very fast as a result of human activities on the Earth.[84] The impacts of climate change are ubiquitous and are projected to grow; "[d]ecisions made today determine risk exposure for current and future generations and will either broaden or limit options to reduce the negative consequences of climate change."[85] More than 40 years ago scientists from 50 nations gathered together at the First World Climate Conference in Geneva.[86] Since then, similar events took place frequently (the 1992 Rio Summit, the 1997 Kyoto Protocol, the 2015 Paris Agreement, etc).[87] Religious organizations also have contributed to the climate change dialogue, highlighting the ethical dimension of the existing problem and addressing the rights of the poor and developing countries.[88] Action to mitigate global climate change must be built upon a foundation of a social and economic justice, as it was declared by the United States Conference of Catholic Bishops.[89] Costs and ways of climate change adaptation and mitigation are expected to vary among developed and developing countries.[90] By 2050, the global economy is projected to increase, as the International Banking Company (HSBC) predicts, but the current practices are far from sustainable.[91]

Climate change is not only an urgent problem, but it is also a fundamental spiritual question concerning social justice and sustainable peace development as well as solidarity among people of various religious backgrounds and different countries. Thus, this global problem must be faced and recognized for future actions and strategies. However, the politics of fear must be replaced with a culture of peace, hope, and compassion, and this urgent problem must be faced with an optimistic attitude and a certain degree of preparedness. According to the World Bank Group Climate Change Action Plan 2016–2020,

> Climate change presents enormous challenges and opportunities for development, making it essential that climate and development be tackled in an integrated way. The world needs to feed nine billion people by 2050, provide affordable energy access to all, and extend housing and services to two billion new urban dwellers—and to do so while minimizing emissions and boosting resilience.

This volume presents a reflection on sustainable development and the proliferation of renewable energy worldwide, in conjunction with such important questions, as social justice and economic growth, providing an interdisciplinary approach to sustainability development. Global climate is projected to continue to change over the twenty-first century and beyond.[92] As the Climate Science Special Report states, "Thousands of studies conducted by researchers around the world have documented changes in surface, atmospheric, and oceanic temperatures; melting glaciers; diminishing snow cover; shrinking sea ice; rising sea levels; ocean acidification; and increasing atmospheric water vapor."[93] There is a great risk of "anticipated changes and impacts, some of which are potentially large and irreversible."[94] Climate change is already transforming the way of life presenting challenges quality of life and to human health, the economy, and the natural systems as well.[95] Risks related to climate change are very difficult to assess; they vary by geographic location and by the vulnerability of people experiencing impacts.[96] Risks are higher for those that are already vulnerable, including low-income communities, children, and the elderly.[97] Human-caused climate change widens the existing social and economic inequality and injustice.

Notes

1. Climate Science Special Report "Physical Drivers of Climate Change," https://science2017.globalchange.gov/chapter/2/ (last accessed 02/18/2020).
2. National Climate Assessment Report, "Climate Change Impacts in the United States," https://nca2014.globalchange.gov/ (last accessed 02/06/2020).
3. The Rio+20 Summit took place as global environmental justice movements.
4. "The Paris Agreement and NDCs," https://unfccc.int/process-and-meetings/the-paris-agreement/nationally-determined-contributions-ndcs#eq-1 (last accessed 02/23/2020).
5. Ibid.
6. "Transforming Our World: The 2030 Agenda for Sustainable Development," https://sustainabledevelopment.un.org/post2015/transformingourworld (last accessed 02/23/2020).
7. Dina Abbott and Gordon Wilson, "Climate Change: Lived Experience, Policy and Public Action," *International Journal of Climate Change Strategies and Management* 6, no. 1 (2014): 5.
8. The Climate Science Special Report, U.S. Global Change Research Program, https://science2017.globalchange.gov/downloads/CSSR2017_FullReport.pdf, 10 (last accessed 02/28/2020).
9. "World Scientists' Warning of a Climate Emergency," https://scientistswarning.forestry.oregonstate.edu/sites/sw/files/climate%20emergency%20Ripple%20et%20al.pdf (last accessed 02/18/2020).
10. Michael B. McElroy, *Energy and Climate: Vision for the Future* (Oxford University Press, 2016), 2.
11. Ibid.
12. Ibid., 11.
13. Ibid., 10.
14. Ibid.
15. Julian Agyeman and Briony Angus, "The Role of Civic Environmentalism in the Pursuit of Sustainable Communities," *Journal of Environmental Planning and Management* 46, no. 3 (2003): 346.
16. "World Scientists' Warning of a Climate Emergency," https://scientistswarning.forestry.oregonstate.edu/sites/sw/files/climate%20emergency%20Ripple%20et%20al.pdf (last accessed 02/18/2020).

17. Michael B. McElroy, *Energy and Climate: Vision for the Future* (Oxford University Press, 2016), 1.
18. The Climate Science Special Report, U.S. Global Change Research Program, https://science2017.globalchange.gov/downloads/CSSR2017_FullReport.pdf, 10 (last accessed 02/28/2020).
19. "World Scientists' Warning of a Climate Emergency," https://science2017.globalchange.gov/downloads/CSSR2017_FullReport.pdf, 29 (last accessed 02/18/2020).
20. Michael B. McElroy, *Energy and Climate: Vision for the Future*, 2.
21. "World Scientists' Warning of a Climate Emergency," https://scientistswarning.forestry.oregonstate.edu/sites/sw/files/climate%20emergency%20Ripple%20et%20al.pdf (last accessed 02/18/2020).
22. "Swallowed by the Sea"—*The New York Times*, https://www.nytimes.com/2018/01/19/opinion/sunday/climate-change-bangladesh.html (last accessed 02/05/2020).
23. Dina Abbott and Gordon Wilson, "Climate Change: Lived Experience, Policy and Public Action," *International Journal of Climate Change Strategies and Management* 6, no. 1 (2014): 8.
24. See Dmitry Kurochkin, Elena V. Shabliy, and Ekundayo Shittu (eds.), *Renewable Energy: International Perspectives* (Palgrave Macmillan, 2019), please see "Conclusion."
25. Martin Robra, "Uppsala Interfaith Climate Manifesto 2008: By Church of Sweden, Editor," *Ecumenical Review* 62, no. 2 (2010): 242.
26. "Pope Francis Joins Climate Fight," United Nations Climate Change, https://unfccc.int/news/pope-francis-releases-encyclical-on-climate-and-environment (last accessed 9/10/2019).
27. Encyclical Letter *LAUDATO SI'* of the Holy Father Francis on Care for Our Common Home, http://w2.vatican.va/content/dam/francesco/pdf/encyclicals/documents/papa-francesco_20150524_enciclica-laudato-si_en.pdf, 3 (last accessed 9/10/2019).
28. Ibid.
29. Ibid.
30. Ministry of the Environment of Sweden, Sweden's Sixth National Communication on Climate Change, https://www.government.se/contentassets/94d274fef8ef470a9b6901421b50d1d1/swedens-sixth-national-communication-on-climate-change-under-the-united-nations-framework-convention-on-climate-change-ds-201411, 4 (last accessed 9/11/2019).
31. Ibid.

32. The Fourth National Climate Assessment, https://nca2018.globalchange.gov/chapter/1/ (last accessed 02/07/2020).
33. Ibid.
34. Ibid.
35. "World Scientists' Warning of a Climate Emergency," https://scientistswarning.forestry.oregonstate.edu/sites/sw/files/climate%20emergency%20Ripple%20et%20al.pdf; "The Ethics of Climate Change," Richard Somerville, 2008, https://e360.yale.edu/features/the_ethics_of_climate_change (last accessed 02/18/2020).
36. Ibid.
37. Ibid.
38. Ibid.
39. United Nations Treaty Collection, Chapter XXVII, "Environment," Paris, 12 December 2015, https://treaties.un.org/pages/ViewDetails.aspx?src=TREATY&mtdsg_no=XXVII-7-d&chapter=27&lang=en; https://climateactiontracker.org/countries/russian-federation/ (last accessed 02/13/2020).
40. Ibid.
41. "Islamic Declaration on Climate Change," https://unfccc.int/news/islamic-declaration-on-climate-change (last accessed 01/11/2019).
42. "Hinduism and Climate Change", April 2013, by Makarand Paranjape, http://fore.yale.edu/news/item/hinduism-and-climate-change/ (last accessed 01/11/2019).

    O! Mother Earth, who has the ocean as clothes
    and mountains and forests on her body,
    who is the wife of Lord Vishnu, I bow to you.
    Please forgive me for touching you with my feet.

43. "Climate Change and Human Rights," Columbia University Law School, Sabin Center for Climate Change Law, https://web.law.columbia.edu/sites/default/files/microsites/climate-change/climate_change_and_human_rights.pdf (last accessed 06/21/2019).
44. Ibid.
45. K. Jenkins, D. Mccauley, R. Heffron, H. Stephan, and R. Rehner, "Energy Justice: A Conceptual Review," *Energy Research & Social Science* 11(C) (2016): 174–182.
46. Ibid.

47. Aladdine Joroff. "Energy Justice: What It Means and How to Integrate It into State Regulation of Electricity Markets." *Environmental Law Reporter* 47, no. 11 (2017): 10933.
48. Benjamin K. Sovacool and Michael H. Dworkin, *Global Energy Justice: Problems, Principles, and Practices* (Cambridge: Cambridge University Press), 5.
49. John Rawls, *A Theory of Justice* (Harvard University Press, 1971), 3.
50. See Dmitry Kurochkin, Elena V. Shabliy, and Ekundayo Shittu (eds.), *Renewable Energy: International Perspectives* (Palgrave Macmillan, 2019), please see Chapter 1, Mishka Lysack, "Economic and Political Foundations of Effective Transition to Renewable Energy: *Ordoliberalism*, Polanyi, and Cities as Hubs for Climate Leadership and Innovation."
51. K. Jenkins, D. Mccauley, R. Heffron, H. Stephan, and R. Rehner, "Energy Justice: A Conceptual Review," *Energy Research & Social Science* 11(C) (2016): 174–182.
52. Dmitry Kurochkin, Elena V. Shabliy, and Ekundayo Shittu (eds.), *Renewable Energy: International Perspectives* (Palgrave Macmillan, 2019), please see Chapter 1, Mishka Lysack.
53. "The Tenth Anniversary of Massachusetts v. EPA," https://www.energycentral.com/c/ec/tenth-anniversary-massachusetts-v-epa.
54. California, Connecticut, Illinois, Maine, Massachusetts, New Jersey, New Mexico, New York, Oregon, Rhode Island, Vermont, and Washington.
55. Supreme Court of the United States, https://www.supremecourt.gov/opinions/06pdf/05-1120.pdf (last accessed 02/18/2020).
56. "Massachusetts v. Environmental Protection Agency," Oyez, https://www.oyez.org/cases/2006/05-1120 (last accessed 10/02/2020).
57. "The Tenth Anniversary of Massachusetts v. EPA," https://www.energycentral.com/c/ec/tenth-anniversary-massachusetts-v-epa (last accessed 02/18/2020).
58. Ibid.
59. "Climate Change in a Nutshell: The Gathering Storm," James Hansen http://www.columbia.edu/~jeh1/mailings/2018/20181206_Nutshell.pdf (last accessed 02/23/2020).
60. Ibid.
61. Ibid.
62. Benjamin K. Sovacool and Michael H. Dworkin, *Global Energy Justice: Problems, Principles, and Practices* (Cambridge: Cambridge University Press), 11.
63. Ibid., 12.

64. "Climate Change Impacts in the United States," http://s3.amazonaws.com/nca2014/low/NCA3_Climate_Change_Impacts_in_the_United%20States_LowRes.pdf?download=1 (last accessed 02/27/2020).
65. "Solar Power Will Kill Coal Faster Than You Think," https://www.cleanenergywire.org/news/nrw-curb-wind-power-germany-criticises-us-russia-sanction-plans/solar-power-will-kill-coal-faster-you-think (last accessed 02/05/2020).
66. Ibid.
67. Ibid.
68. Ibid.
69. Ibid.
70. "ЕВРОСОЛАР Россия," EurosolarRussia, International Congress REENCON-XXI "Renewable Energy-XXI Century: Energy and Economic Efficiency," http://www.eurosolarrussia.org/en/meropriyatiya/forumy_konferentsii_kruglye_stoly_delovye_zavtraki/_reencon_xxi_xxi_/ (last accessed 02/18/2020).
71. Common Wealth of Virginia Office of the Governor Executive Order, November Forty-Three (2019), https://www.governor.virginia.gov/media/governorvirginiagov/executive-actions/EO-43-Expanding-Access-to-Clean-Energy-and-Growing-the-Clean-Energy-Jobs-of-the-Future.pdf, 2–3 (last accessed 02/18/2020).
72. Ibid.
73. Ibid.
74. Ibid.
75. Ibid., 4 (My emphasis).
76. The Virginia Clean Economy Act, https://www.vacleaneconomy.org/ (last accessed 02/18/2020).
77. "100% Commitments in Cities, Counties, & States," The Sierra Club, https://www.sierraclub.org/ready-for-100/commitments; "More Than 250 US Mayors Aim at 100% Renewable Energy by 2035," https://unfccc.int/news/more-than-250-us-mayors-aim-at-100-renewable-energy-by-2035 (last accessed 02/18/2020).
78. Ibid.
79. Ibid.
80. Ibid.
81. Heike Köckler, Severiné Deguen, Andrea Ranzi, Gordon Walker, Anders Melin, Ryan Holifield, and Jayajit Chakraborty, "Environmental Justice in Western Europe," *Routledge Handbook of Environmental Justice*, 2017.

82. Benjamin K. Sovacool and Michael H. Dworkin, *Global Energy Justice: Problems, Principles, and Practices* (Cambridge: Cambridge University Press), 1.
83. K. Jenkins, D. Mccauley, R. Heffron, H. Stephan, and R. Rehner, "Energy Justice: A Conceptual Review," *Energy Research & Social Science* 11(C) (2016): 174–182.
84. The Fourth National Climate Assessment, https://nca2018.globalchange.gov/chapter/1/ (last accessed 02/07/2020).
85. Ibid.
86. "World Scientists' Warning of a Climate Emergency," https://scientistswarning.forestry.oregonstate.edu/sites/sw/files/climate%20emergency%20Ripple%20et%20al.pdf (last accessed 02/18/2020).
87. Ibid.
88. "The Ethics of Climate Change," https://e360.yale.edu/features/the_ethics_of_climate_change (last accessed 02/18/2020).
89. Ibid.
90. Ibid.
91. Michael B. McElroy, *Energy and Climate: Vision for the Future* (Oxford University Press, 2016), 7. See also "The World in 2050: Quantifying the Shift in the Global Economy," Global Economics, HSBC Global Research, Global Economics 2011, https://www.hsbc.ca/1/PA_ES_Content_Mgmt/content/canada4/pdfs/business/hsbc-bwob-theworldin2050-en.pdf (last accessed 02/18/2020).
92. The Climate Science Special Report, U.S. Global Change Research Program, https://science2017.globalchange.gov/downloads/CSSR2017_FullReport.pdf, 15 (last accessed 02/18/2020).
93. Ibid., 10.
94. Ibid., 11.
95. The Fourth National Climate Assessment, https://nca2018.globalchange.gov/chapter/1/ (last accessed 02/07/2020).
96. Ibid.
97. Ibid.

Bibliography

Abbott, D., & Wilson, G. (2014). Climate Change: Lived Experience, Policy and Public Action. *International Journal of Climate Change Strategies and Management*, 6(1), 5–18.

Agyeman, J., & Angus, B. (2003). The Role of Civic Environmentalism in the Pursuit of Sustainable Communities. *Journal of Environmental Planning and Management*, 46(3), 345–363.

"Climate Change and Human Rights," Columbia University Law School, Sabin Center for Climate Change Law, https://web.law.columbia.edu/sites/default/files/microsites/climate-change/climate_change_and_human_rights.pdf, last accessed 06/21/2019.

Common Wealth of Virginia Office of the Governor Executive Order, November Forty-Three (2019), https://www.governor.virginia.gov/media/governorvirginiagov/executive-actions/EO-43-Expanding-Access-to-Clean-Energy-and-Growing-the-Clean-Energy-Jobs-of-the-Future.pdf, last accessed 02/18/2020.

Encyclical Letter *LAUDATO SI'* of the Holy Father Francis on Care for Our Common Home, http://w2.vatican.va/content/dam/francesco/pdf/encyclicals/documents/papa-francesco_20150524_enciclica-laudato-si_en.pdf, last accessed 02/28/2020.

The Environmental Group Sierra Club, "100% Commitments in Cities, Counties, & States," https://www.sierraclub.org/ready-for-100/commitments, last accessed 02/2/2020.

The Fourth National Climate Assessment, https://nca2018.globalchange.gov/chapter/1/, last accessed 02/07/2020.

"Hinduism and Climate Change," April 2013, by Makarand Paranjape, http://fore.yale.edu/news/item/hinduism-and-climate-change, last accessed 01/11/2019.

Holifield, R., Chakraborty, J., & Walker, G. *The Routledge Handbook of Environmental Justice* (Routledge Handbooks). London; New York: Routledge, Taylor & Francis Group, 2018.

"Islamic Declaration on Climate Change," https://unfccc.int/news/islamic-declaration-on-climate-change, last accessed 01/11/2019.

Jenkins, K., Mccauley, D., Heffron, R., Stephan, H., & Rehner, R. (2016). Energy Justice: A Conceptual Review. *Energy Research & Social Science*, 11(C), 174–182.

Joroff, Aladdine. (2017). Energy Justice: What It Means and How to Integrate It into State Regulation of Electricity Markets. *Environmental Law Reporter*, 47(11), 10933.

Köckler, Heike, Deguen, Severiné, Ranzi, Andrea, Walker, Gordon, Melin, Anders, Holifield, Ryan, & Chakraborty, Jayajit. (2017). Environmental Justice in Western Europe. In *Routledge Handbook of Environmental Justice*. New York: Routledge.

Kurochkin, Dmitry, Shabliy, Elena V., & Shittu, Ekundayo. *Renewable Energy: International Perspectives on Sustainability*. Cham: Palgrave Macmillan, 2019.

Lysack, M. *Economic and Political Foundations of Effective Transition to Renewable Energy: Ordoliberalism, Polanyi, and Cities as Hubs for Climate Leadership and Innovation*. Cham: Springer International Publishing, 2019.

McElroy, Michael B. *Energy and Climate: Vision for the Future*. New York: Oxford University Press, 2016.

Ministry of the Environment of Sweden, Sweden's Sixth National Communication on Climate Change, https://www.government.se/contentassets/94d274fef8ef470a9b6901421b50d1d1/swedens-sixth-national-communication-on-climate-change-under-the-united-nations-framework-convention-on-climate-change-ds-201411, last accessed 9/11/2019.

"More Than 250 US Mayors Aim at 100% Renewable Energy by 2035," https://unfccc.int/news/more-than-250-us-mayors-aim-at-100-renewable-energy-by-2035, last accessed 02/18/2020.

National Climate Assessment Report, "Climate Change Impacts in the United States," https://nca2014.globalchange.gov/, last accessed 02/06/2020.

"The Paris Agreement and NDCs," https://unfccc.int/process-and-meetings/the-paris-agreement/nationally-determined-contributions-ndcs#eq-1, last accessed 02/23/2020.

"Physical Drivers of Climate Change," Climate Science Special Report, https://science2017.globalchange.gov/chapter/2/, last accessed 02/18/2020.

"Pope Francis Joins Climate Fight," United Nations Climate Change, https://unfccc.int/news/pope-francis-releases-encyclical-on-climate-and-environment, last accessed 9/10/2019.

Robra, M. (2010). Uppsala Interfaith Climate Manifesto 2008. By Church of Sweden, editor. *Ecumenical Review*, 62(2), 242.

Somerville, Richard. "The Ethics of Climate Change," Richard Somerville, 2008, https://e360.yale.edu/features/the_ethics_of_climate_change, last accessed 02/18/2020.

Sovacool, B., & Dworkin, M. *Global Energy Justice: Problems, Principles, and Practices*. Cambridge: Cambridge University Press, 2014.

"Swallowed by the Sea"—*The New York Times*, https://www.nytimes.com/2018/01/19/opinion/sunday/climate-change-bangladesh.html, last accessed 02/05/2020.

"Transforming Our World: The 2030 Agenda for Sustainable Development," https://sustainabledevelopment.un.org/post2015/transformingourworld, last accessed 02/23/2020.

United Nations Treaty Collection, Chapter XXVII, "Environment," https://treaties.un.org/pages/ViewDetails.aspx?src=TREATY&mtdsg_no=XXVII-7-d&chapter=27&lang=en, Paris, 12 December 2015, last accessed 02/28/2020.

The Virginia Clean Economy Act, https://www.vacleaneconomy.org/, last accessed 02/18/2020.

Walker, G. *Environmental Justice: Concepts, Evidence and Politics*. London: Routledge, 2011.

"The World in 2050: Quantifying the Shift in the Global Economy," Global Economics, HSBC Global Research, Global Economics 2011, https://www.hsbc.ca/1/PA_ES_Content_Mgmt/content/canada4/pdfs/business/hsbc-bwob-theworldin2050-en.pdf, last accessed 02/18/2020.

"World Scientists' Warning of a Climate Emergency," https://scientistswarning.forestry.oregonstate.edu/sites/sw/files/climate%20emergency%20Ripple%20et%20al.pdf, last accessed 02/18/2020.

2

Public Participation and Intra-Actions in the Swedish Energy Transition

Annika Skoglund and Jessica García-Terán

Introduction

The Swedish government has over the years been very proactive in their attempts to create citizen awareness of human-induced climate change (Uggla 2008) and Sweden is often pointed out internationally as a vanguard of climate change mitigation and adaptation (Anshelm and Hultman 2014). Although sustainability has been taken seriously for a long time by the Swedish state, some businesses, and many citizens (Gustavsson et al. 2009), repeated calls for more action nevertheless point to the difficulty of reaching beyond mere talk (Langlais 2009). Hence, efforts made by the Swedish state, and its proponents, to conquer the moral high ground, have been strongly questioned by climate

A. Skoglund (✉) · J. García-Terán
Department of Civil and Industrial Engineering,
Uppsala University, Uppsala, Sweden
e-mail: annika.skoglund@angstrom.uu.se

J. García-Terán
e-mail: jessica.garcia@angstrom.uu.se

E. V. Shabliy et al. (eds.), *Discourses on Sustainability*,
https://doi.org/10.1007/978-3-030-53121-8_2

activists (Thunberg 2019). At the same time as the Swedish sustainability discourse is going strong, climate actions are thus suggested to be weak. The labor union for academics in the natural sciences, *Naturvetarna*, has, for example, emphasized this view. The union rejects strike actions and emotional expressions in a counter to the passionate and occasionally angry Swedish school-striking youth climate activist Greta Thunberg, and advise their members to take more hands-on and direct actions (Naturvetarna 2019). The work done by their mature members will lead to the awaited innovative solutions to human-induced climate change, they rhetorically propose.

Complaints about emotional responses to change, in comparison to the celebration of acts of control, follow a common (gendered) criticism that extends beyond the type of changes that are described via our cultural understanding of the climate. Both 'change' and the 'climate' are widely used generic terms, and put together, these two terms—'climate change'—become extremely pervasive and powerful. Not only does the current discourse on climate change thrive on a tone of danger and catastrophe (Hoggett 2011), as Swedish media testify (Höijer 2010), but 'throughout the human experience of realized climate, and in anticipation of portended climates, runs a thread of anxiety and fear', and also 'mythologizing and taming' (Hulme 2008: 5–6). Perhaps of specific importance for people surviving in the Northern hemisphere, 'taming' has developed into modern forms of control, either of the climate itself, or of how citizens are to behave differently in relation to a changed climate. Hence, in Sweden, market-driven solutions and innovations have prospered hand in hand with ecological modernization and participative democracy programs, or 'deliberation', promoted by state agencies (Hildingsson 2010), and the most prominent area that encompasses both control of the climate, and of the emotional citizen, is the transition to a sustainable, or so-called 'smart' or 'intelligent', energy system (Brange et al. 2016; Öhrlund et al. 2019).

In this book chapter, we will further investigate how the local government, businesses, and citizens are connected to take action against human-induced climate change in a specific case: the renovation of a bridge in the city of Uppsala. During the renovation, an attempt was made to include a new innovation of a renewable energy technology, developed at Uppsala University, in the new construction planned by the municipality of Uppsala. While the researchers, or the so-called 'academic entrepreneurs' (Erdős and Varga 2012), focused on basic 'blue sky' research and the creation of applications, Uppsala municipality wished to respond to climate change as part of the city brand and their stimulation of citizen inclusion and voicing. The two actors, the academic entrepreneurs, and the public servants, therefore met on common ground, the Triple and Quadruple Helix models (Leydesdorff 2010; Etzkowitz 2003), occasionally supported by 'the Swedish Governmental Agency for Innovation Systems' (VINNOVA).

As we will illustrate in more detail, the renovation project first included three spheres: industry, state, and the university. During the planning and implementation, however, a fourth sphere was added: civil society (Carayannis and Campbell 2009, 2011b). Aside from this increased attention given to public participation and bottom-up movements, our explorative and ethnographic study of the renovation project showed that the renewable energy technology played a considerable role as a complementary agent, acting through negotiations full of material and social concerns (Latour and Woolgar 1986). To better understand the process of materialization in front of us, it thus seemed analytically important to bring the technology per se directly into the model, and include it in a sphere of its own. Analytically, we thus complemented the original model by constituting a new actor in a fifth sphere. Even though a fifth sphere has been introduced to the innovation system before, this has mainly been done to represent the environmental conditions in a specific region, or 'nature' as an actor to stimulate innovation in line with the sustainable development agenda (Carayannis et al. 2012; Carayannis and Campbell 2010; Grundel and Dahlström 2016; Carayannis and Rakhmatullin 2014) (see Fig. 2.1a–c). In contrast, we thus identified a need to better understand how the technology sometimes became an active part, and sometimes an inactive part, of the synergies created in between ecological modernization and democratic deliberation.

Since very little theoretical attention has been paid to interactions with non-human agents in innovation models, there is a need to expand the theoretical territory with which to develop these quite ideal and fixed models. We, therefore, turn to posthuman thinking with the aim

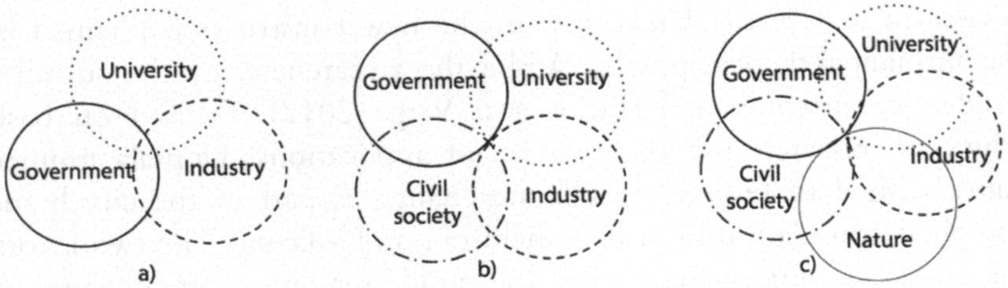

Fig. 2.1a–c From the Triple Helix to the Quadruple Helix, to a fifth helix model. The added sphere affirms a posthuman turn and constitutes nature (or technology) as an intra-actor

of investigating 'interactions' differently, as 'intra-actions' (Barad 2007). This is a concept that pinpoints the fluidity of matter, defined in opposition to a fixed substance, assumed to be 'a substance in its intra-active becoming—not a thing but a doing, a congealing of agency' (Barad 2007: 151). We thus contribute to studies on the sustainability discourse from a posthuman materialist angle, with an emphasis on how one of the solutions to climate change, i.e., renewable energy technology, intra-acts with citizens, academic entrepreneurs, and governmental agencies.

The chapter is organized as follows. After an introduction to studies of 'interactions' in innovation systems, we provide a theoretical overview of how we may shift conceptually from interactions to intra-actions. Thereafter follows a more explicit method section, explaining three useful analytical tools with which to study the emerging process of materialization and public participation: intra-actions, noticings, and boundary re-constructions. We then briefly exemplify what the addition of a fifth sphere does in practice, by applying our proposed analytical tools to a specific case. We end with a discussion about our advancement of the helix model, which includes both a more processual as well as posthuman attempt at understanding innovation and commercialization in the energy transition.

'Interactions' Between Human Actors in the Innovation System

Studies of the Triple Helix (TH) model in various innovation systems focus their analyzes on how governments, universities, and industries interact in efforts to commercialize technology (Etzkowitz and Zhou 2017). With ambitions to both strengthen democracy and commercialization processes, recent advancements of the TH model have added interactions with a fourth sphere, civil society (Carayannis and Campbell 2009, 2011b). Here, 'interactions' are what supposedly drive collaborations into more or less successful 'fluid and heterogeneous innovation networks' (Carayannis and Campbell 2011b: 342). Seemingly, these analyzes of helix innovation processes thus follow a methodology that assumes and encircles human actors as the main agentic and interacting subjects. It is a methodology that, similarly to traditional conceptualizations of communication, refers to a structural landscape where people speak matter into being (e.g., see Packer and Crofts Wiley 2012).

Tellingly, it is mainly 'interactions' between human actors that are pinpointed as crucial. It is these that are conceived to either facilitate or hinder technical change. Even though this realization about interactions has become increasingly evident in the literature, few studies focus explicitly on interactions, and there is no consensus or even a thorough elaboration on their ontological status. Some studies are based on a macro perspective, discussing overarching interactions between spheres, while others use a meso perspective, discussing how local actors interact with larger institutions, while a third perspective focuses on an interest in the interactions between individuals who work in the innovation system.

Lee and Kim (2016), for instance, investigate R&D networks in national R&D programs in South Korea's government policies and suggest that policymakers continuously engage in feedback loops with triple helix institutional actors, concluding that governments should encourage and not force or direct R&D networks since this can obstruct the dynamic interactions resulting from R&D networks. Flexible interactions are thus suggested to provide access to novel information from other actors, where such interactions should be stimulated so as to inspire interaction between new actors through the support of network and coordination capabilities. Interactions are furthermore positioned in relation to big data systems for R&D networking support, where information and knowledge should be shared among the actors for new co-operation opportunities. Hence, 'interactions' are loosely coupled to 'networking' and a successful sharing between various actors in the network.

In comparison, Sarpong et al. (2017) use a case study approach to look at organizing practices of innovation in Malaysia. They argue that for a successful transition to a triple helix model of innovation it is not enough to have interdependent relationships and interactions between the spheres, but instead there is a need to look at the reconfiguration of the actor's doings, routines, and organizing practices, as these elements are what is pointed out as important to understand for an enhancement of the innovation outcomes. However, organizing practices are neither conceived in a process of becoming, nor something embedded in organizations, but characterized as a collective form of agreements, which 'have a history, are flexible, and are in a constant flux of transformation' (Sarpong et al. 2017: 144). These collective agreements are expressed by the 'doings' and 'sayings' that interconnect the spheres in actual situations, driving the transition process in a specific direction. Hence, analytical focus is turned to the use of language, patterned activities, and practices, said to explain the transition process. By empirically selecting 'strategic actors', rather than any interactions that unfold, the authors are curious about 'how the recent emphasis on collaborations have [sic] impacted on their situated practice' (ibid.: 146). Three organizing practices are identified: (1) research capabilities and partnerships, (2) quantification of knowledge, and (3) collective entrepreneurship (ibid.: 147). These practices are fleshed out by how everyday routines are described to be involved in the strategic relationships and collaborative processes that are created between the TH actors. Hence, the authors get closer to a processual perspective in the end and invite scholars to persist in exploring the contingency role of organizing practices and how they may contribute to a smooth transition of countries to an idealized innovation system.

Another study with a focus on the creation of Google follows a processual approach to how interactions unfold between actors in the development of an 'integrated virtual library' (Steiber and Alänge 2013). In this case, interactions emerged as researchers, professors, and supervisors became involved with private companies and funding agencies, suggesting that it is the interactions among various stakeholders that shaped the story of Google. Patent applications, business competitions, business angels, and later on investors, all contribute to the story of how Google was created. The authors especially highlight the role of the university, as the provider of a fruitful environment for the initial idea, with emphasis on how the interactions are kept alive after the start-up of the business by joint research programs. Interactions are also understood in the form of 'dialogue', and how such is maintained on a continuous basis with the government, much due to the steering of market regulations that affect Google. The dynamism of all the interactions is highlighted as a balance is sought between the different spheres over time, suggesting that it is crucial 'to understand and even orchestrate the interactions between the helices in order… to create a good environment' (ibid.: 598). Interactions are also distinguished into formal and informal, and these are discussed in relation to Google's formation and growth. The authors argue that in order to understand TH interactions, one needs to pose 'how questions', and look at the circumstances even 'before the entrepreneurs came on the scene' (ibid.: 597). They argue that pre-business establishment interactions are important to address since these set the scene for the business via the universities and the government.

A complementary processual approach to interactions is suggested by Lindgren and Packendorff (2002), who study innovative social processes as 'organic processes, as an open-ended series of events in which people create things together'. Innovative processes are thus conceived to emerge from the interaction between individual actors (ibid.), and by looking at innovation processes as socially constructed, they suggest that analyzes of TH interactions could focus on how opportunities are recognized and legitimized, and what drives social processes forward. In this way, studies can take into consideration how innovation processes unfold, for example, how ideas emerge and how interactions have an effect on the future path. Instead of an analytical focus on the individual actors, it is the innovative social process that is of main interest. Following Lindgren and Packendorff (2002), there are thus efforts made to focus on how interactions can be studied differently, as less causal and more emergent.

This epistemological position of interactions in innovation models combined with a deliberative agenda have recently been challenged with the development of a fifth sphere, which includes the natural environment to encompass the sustainability agenda (see Fig. 2.1a–c). With increasing awareness of global warming and the climate emergency (Wittneben et al. 2012; Newell 2020), the fifth sphere encircles 'nature' to bring it in as a central component for innovation (Carayannis 2012; Peris-Ortiz et al. 2016). Global warming is not to be regarded as a 'challenge', but an 'opportunity', much in line with the agenda on ecological modernization (Carayannis et al. 2012) as well as sustainable development (Grundel and Dahlström 2016). It, therefore, differs from the other spheres since it attempts to grant some agency to non-human things. In comparison to the other four spheres that consist of human actors, the fifth sphere affirms a posthuman turn and constitutes nature as an actor in itself. At least this is the case as long as the fifth sphere is not studied by an analysis of environmental movements and Green NGOs that take it upon themselves to speak on behalf of nature. These organizations and environmental concerns would then belong to the sphere of civil society (cf. Hock Heng et al. 2012). Hence, even if nature is granted some sort of 'agency' in the fifth sphere, how this agency stands in relation to the overall model is still largely underdeveloped.

Furthermore, even though a fifth sphere has been added, the model mainly refers to linear time, where an invention moves metaphorically helix-wise along a time axis to reach a stage of innovation, and hopefully, later commercialization (Carayannis and Campbell 2011a: 339) (see Fig. 2.2). Since conceptualizations of the helix model have, to a very limited extent, included sociomaterial ontologies, how the model supports a process of materialization is also undertheorized. There is thus a need to explore the methodological possibilities for studying the implementation of helix models processually in the human–non-human nexus. This is of specific importance in the attempted transition to a sustainable energy system, which seeks to place the human within a socio-ecological-technical system. Hence, while the fifth sphere previously has included

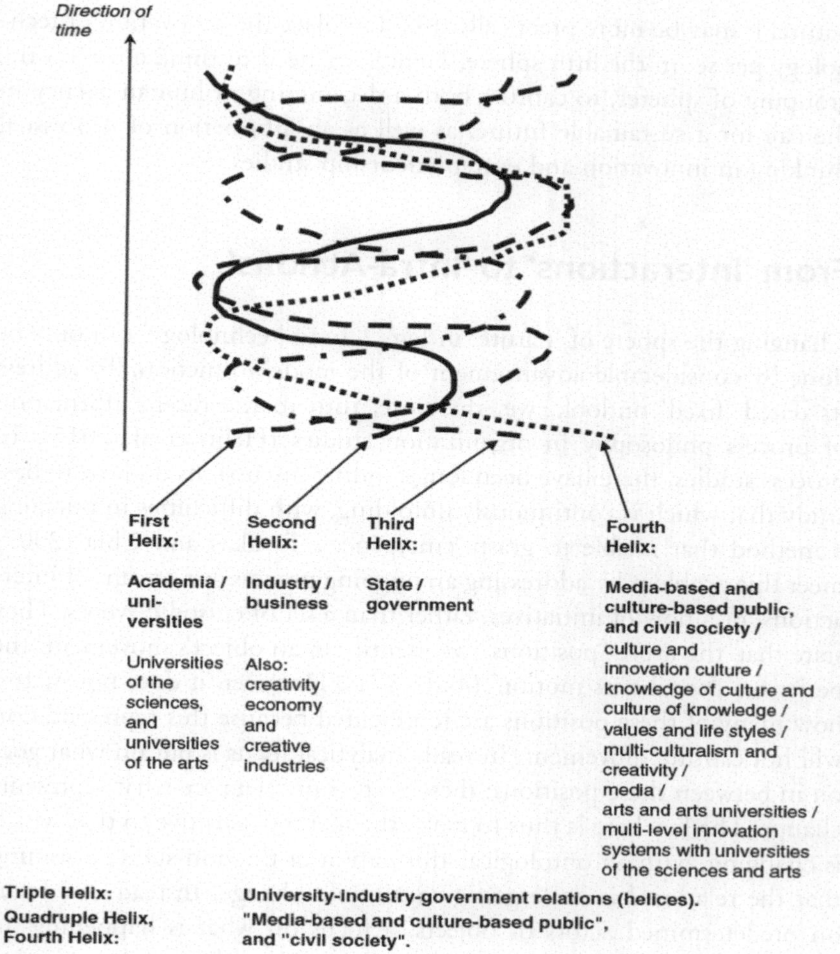

Fig. 2.2 The conceptualization of the Quadruple Helix model (*Notes* The interwoven DNA-type helix illustrates four actors interacting in the innovation system. *Source* Carayannis and Campbell 2011a: 339)

nature, it may be more practically useful to place the innovation or technology per se, in the fifth sphere. Hence, we need to think anew on the grouping of spheres, to capture both a decentering of human agency in the call for a sustainable future, as well as an affirmation of processual thinking in innovation and entrepreneurship studies.

From 'Interactions' to 'Intra-Actions'

Changing the sphere of 'nature' in Fig. 2.1c to 'technology' can only be done by considerable advancement of the model's function. To address its dated 'fixed' outlook, we, therefore, turn to the recent affirmation of process philosophy in organization studies (Helin et al. 2015). In process studies, there have been longstanding discussions on how to best study that which is continuously unfolding, with difficulties in pursuing a method that is able to grasp 'emergence'. Tsoukas and Chia (2002) meet this problem by addressing an ongoing process as a stream of interactions, or a flow of initiatives, rather than a set of episodic events. They state that 'the more "positions" we identify in an object's movement, the better we describe its motion' (ibid.: 571). However, it does not matter how many of these positions are represented because this representation will not capture movement. Instead, analytical focus is put on what goes on in between these positions; these vectors or relata are what represents change. The key here is thus to make the method sensitive to that which is changing, with an ontological affirmation of the non-static, assuming that the relational is undergoing continuous change. Instead of a focus on predetermined actors or objects, a focus on what is happening 'in between' makes it possible to appreciate movement, flow, and continuity. The authors suggest that such a production of knowledge about processes happens via 'knowing from within' (ibid.: 571).

Likewise, Shotter (2006) has emphasized that a processual view requires an understanding 'from within'. This means that the researchers should involve themselves with the process itself and apply an undergirding awareness enmeshed in the flow and fluidity. Shotter (ibid: 585) quotes Bergson (1911: 343–344) to clarify that '…it is the flow of time, it is the very flux of the real that we should be trying to follow…'. Nevertheless, a far too common approach in studies of processes has been to study them 'from the outside', as something happening over there or something that has happened, illustrated retrospectively in a quite linear fashion. The problem with such an observational and analytical approach is that fluidity and dynamism are downplayed. By instead applying a subsidiary awareness, the researcher can seek to experience the process as it is flowing and unfolding. This requires thinking about and engaging with a particular 'becoming' (ibid.).

Methodological discussions about how to apply process philosophy thrive on the prevalent difficulties in concretizing a research agenda that successfully secures that it really is a 'becoming' that is explained and made sense of. With different means to counter the conventional observational power of the human over objects, things, or artifacts, such an agenda mainly rests in attacks on the modern sovereign subject, objectivity, as well as the linguistic turn. In opposition to these, researchers are told to think and pursue research in a very different way, more posthumanly. For example, as a responsive understanding of 'living forms' (Shotter 2006: 589) and the dialogical possibility between living bodies that interact with their surroundings in a spontaneous manner. As such, in the study of processes, it is suggested to include events occurring within meetings, between ourselves and the others and the surroundings, 'events internally related to such meetings' (ibid.: 594). Briefly put, in the study of processes with the help of a turn to 'living forms', changes are viewed as unexpected and unpredictable, made available to us if we are involved in a 'dynamic interactive, expressive-responsive relation with the others and othernesses in our surroundings' (ibid.: 599). Similarly, studies of entrepreneurial processes emphasize a focus on events, together with processual language, to concretize a processual approach (Steyaert 1997). This means an increased application of terminology that describes the movement, such as emergence, evolution, and variation. These terms then influence the way in which researchers are able to think about a process of materialization. To better equip researchers with the ability to actually speak about something that unfolds or is happening, it is additionally suggested to be analytically attentive to dilemmas and interactions, where the 'consecutive stream of events and interactions' is central to developing a processual view (ibid.: 19).

Advancement of the posthuman turn in process studies has nevertheless left interactions behind to study discursive-material 'intra-actions', underpinned by 'performativity' rather than essentialistic psychological approaches (Barad 2003; Ringrose and Rawlings 2015; Iedema 2007). According to Barad, matter should not be neglected, but be allowed to play the role as an active participant, one that contributes to 'the world's becoming, in its ongoing "intra-activity"' (Barad 2003: 803). Matter is thus defined in opposition to a fixed substance and is instead assumed to be 'a substance in its intra-active becoming—not a thing but a doing, a congealing of agency' (Barad 2007: 151).

In comparison to Shotter's background in psychology and mathematics, Karen Barad has a background in physics, and perhaps of significance for the development of their respective processual perspectives, Barad is less interested in living forms and interactions, and more interested in understanding how 'phenomena' (as specifically defined) unfold and develop certain physical properties. She suggests that 'an agential realist elaboration of performativity' should focus on how matter becomes the opposite to the anthropocentric view of cutting it off from relations—'an active participant in the world's becoming, in its ongoing intra-activity' (Barad 2007:136). The ontological units are thereby not independent objects but phenomena that go beyond the merger of the observer and observed to reach an *'ontological inseparability/entanglement of intra-acting "agencies"'* (Barad 2007: 139 italics in original). For her, there is an inseparability between intra-acting agents, which has ontological consequences for how objects/subjects are entangled. These are seen as parts of wider material practices that accomplish 'an agential cut' between subject/object, in contrast to the Cartesian cut, which assumes from the start that this separation is naturally existing (Barad 2007: 140). Hence, the study of phenomena is essentially a study of boundary work—how 'differential boundaries between humans and nonhumans, culture and nature, science and the social, are constituted' (ibid.). Still, there is tension between how matter is considered 'real' and how signifying practices are considered to create realities; a tension solved by positioning 'reality' within intra-actions: 'Reality is composed not of things-in-themselves or things-behind-phenomena but of things-in-phenomena' (ibid.). Furthermore, it is intra-actions that are driving how 'relata-within-phenomena emerge', which has an effect on how causality is viewed in processual analyzes, and it is here that we find intra-actions to be a fruitful conceptual shift for the study of how implementations of the helix model emerges and evolves, and how synergies between democracy, economy, and technology contribute to or thwart a process of materialization of a more sustainable world.

An interest in semantics is not confined to words, but material-discursive practices that are performative of boundaries. A useful analytical tool for Barad's study of phenomena is thereby set out to be 'apparatuses', defined to be 'specific material reconfigurings of the world that not merely emerge in time but iteratively reconfigure spacetimematter as part of the ongoing dynamism of becoming' (Barad 2007: 140). In analogy of laboratory experimental setups, Barad argues that apparatuses are not passive instruments for observation, but active and productive of the phenomena. By extension, that which emerges is not to be understood in relation to linear time, i.e., a causal following of variables that are measured over time (as in the natural sciences), but conceived in a more 'dynamic' way, as the mutual constitution of space, time, and matter. In effect, this makes intra-actions an analytical locus for the 'nonarbitrary, nondeterministic causal enactments through which matter-in-the-process-of-becoming is iteratively enfolded into its ongoing differential materialization' (Barad 2007: 140). This is conclusively the key to Barad's posthumanism—an emphasis on materiality as playing an active part in materialization—it is a performative agent where matter is a doing and not a thing (Barad 2007: 183).

By applying Barad's framework, Ringrose and Rawlings (2015) illustrate how reality emerges via non-human agentic matter in the form of time, objects, and space. Hence, discourse is assumed to be a category belonging to the non-material and human side, while matter is kept on the other side, the non-human. With this move of thought, intra-action gains an epistemological position in between these sides, as bodies, discourses, and context-specific material agents relate to each other in performative ways, or as Iedema (2007) emphasizes, meaning does not arise out of distinct moments, but emerges continuously, whereby a mobilization of 'the concept of the intra-act is to underscore that as humans we are not confronted with a fully completed world out there (or in the head) waiting to be labelled' (Iedema 2007: 941). Intra-actions thus make it possible to think differently about an assembly of actors, for example in the helix innovation model, shaped as posthuman hybrids, not to be spoken about as either objects or subjects.

Consequently, the epistemological position of intra-actions brings agency to the forefront by ascribing it to the relations between actors and matter, or 'spacetimematterings'. Similar to other attempts to leave a human-centered view behind, the non-human, more than the human or posthuman, thereby receives increased analytic attention. Barad's basic assumption is thus that there is no core from which the agency stems, rather 'agencies are dependent on their mutual inextricability', which is fundamental for the creation of fluidity (Nyberg 2009: 1184). It is here that we have a clear methodological difference between ANT (Actor-Network-Theory) and agential realism, in that the latter focuses on processual entanglement on the ground, and not on how a network has unfolded by tracing agency in retrospect (Nyberg 2009). Nyberg (ibid: 1193) thus ends by arguing that an 'assembly of machines, people and the rest into cyborgs' is continuously produced, where meaning can only be understood by a study of the 'contextual performance of the actors', seen as non-determinate categories mobilized via intra-actions (ibid.: 1195).

Analyzing Emergence—Intra-Actions, Noticings, and Boundary Re-constructions

In line with this summary of how intra-actions have been developed and studied by others, we wish to point out certain analytical tools useful for understanding the helix innovation model processually. First, we wish to stress, how Barad's advancement of spacetimematterings cuts out the unimportant, or what is called 'differential matterings' (Ringrose and Rawlings 2015: 91), something that Shotter (2013) discusses further in a chapter focusing on 'intra-thinking'; i.e., a repetition of the methodological problem of how to equip researchers with analytical tools apt for an embracement of processual flux. The answer to this methodological problem is exactly the conceptual shift we have proposed, leaving interactions behind to rather look at intra-actions (ibid.). It is a move away from fixed and static 'things-in-themselves' toward the possibility of seeing all things as 'doings' in boundaryless fluid processes of becoming that inevitably enact new boundaries. Shotter places Barad's 'intra-actions' next to those of Bakhtin, James, and Wittgenstein, since these thinkers contributed to the merger of the social and material in their own ways. He does so by keeping a polite distance from Barad's physics analogies, for example, 'diffractions', to rather grasp how human agency is embedded in flowing processes and 'movements of bodily feeling, occurring in our material intra-actions' (Shotter 2013: 3). Important in the study of the processually emergent then, this leads to how we can get analytically closer to 'emergence'. The focus on intra-action, instead of interaction, makes it possible to analyze 'emergence' without relying on Cartesian rationality of causality. That is, even if effects occur due to a certain component, it is not always possible to trace a causality, rather the effect should be called 'an emergent' since '[i]t arises out of the combined agencies, but in a form which does not display the agents in action' (Lewes 1875, pp. 368–369 cited in Shotter 2013). The analytical focus is neither on actors in themselves, nor the fluidity of waves, but strands of flowing that happen via intra-actions between actors, whereby events unfold and non-determinate organizing emerges.

By understanding emergence, we can thus grasp how 'innovative change' unfolds (Shotter 2013: 5). That is, innovative change cannot be planned and executed according to plans, where everyone acts in accordance with preset paths. Innovation is rather happening in the non-determinate friction created when agencies intermingle. Something that we can only study if we expose ourselves, as embodied beings, to a 'flow of experience' and 'an unceasing flow of language intra-twined activities' (Shotter 2013: 6, 7). Hence, in Shotter's posthumanism, a human who can experience the processual becoming is still highly relevant for the analysis. Methodologically, a study of 'emergence' also shares affinities with studies of prefiguration, constituted by corporeal performativity attached to speech in actuality (see further Felman 2002). Emergence, thus, dissolves the separation of talk and action, but also of 'inside' and 'outside', the objective and the subjective, the material and the mental, as Shotter illustrates (2013: 8):

> They [utterances] are neither 'inside' people, but nor are they 'outside' them; they are located in that space where inside and outside are one; nor is there a separate before and after (Bergson), neither an agent nor an effect, but only a meaningful, 'enduring' intra-acting whole which cannot divide itself into separable inter-acting parts.

In addition, Shotter (2013) proposes a way to actually study emergence by leaving known conceptualizations, culturally determined ideas and ways of seeing behind, and instead turn to 'noticings', where we are to let ourselves be struck by events, sense a unique 'unitary whole' or an error. A noticing can also be sensed in the formation of community, public life or solidarity, or in the unsaid and the left out. Noticings, furthermore, can be experienced when people align in mutual understandings, or when they have not yet done so. These 'noticings' are thereby guidance for how to get closer to the spontaneous, what is just happening, and that which is unfamiliarly emerging via intra-actions.

Holford (2013) equally turns to agential realism, but to advance the concept of boundary object in a move away from the object viewed as a thing, to leave the subject/object duality behind. This is important, he argues, if we wish to go beyond human–object interactions, and instead study human–object intra-actions, which are closer to the dynamic that is at play in knowledge formations. The empirical curiosity investigated has to do with the dissemination of knowledge, and how boundary objects have been thought to be useful for effective knowledge sharing. Since this view holds dear that the subject has an interpretive role, the object becomes a separate entity, 'either being static, or as a changing entity in which we focus only on its static end-result' (Holford 2013: 16). As a contrast to this artifact-centered view, a more process-oriented perspective is proposed. In a complaint about the search for objective truth claims, Holford re-makes boundary objects and proposes 'that they are shaped by actor-subjects (via enactment), just as they also shape actor-subjects (via the subject's interpretative schemes)' (Holford 2013: 19). Since this happens in a performative process of becoming, 'boundary (re)constructions' are launched as an alternative to boundary objects, to enable focus on the agency within a practice, the 'construction' as it happens, instead of the artifact, the 'construct'.

This summary of how interactions in the helix model can be exchanged for a study of 'intra-actions' clearly shows a re-positioning of 'agency' in the grouping of spheres. If the notion of 'interaction' in the original helix model was underpinned by an assumption that predetermined how spheres of actors should be related to other actors, we thus suggest a loosening of the predetermined boundaries and properties of actors according to Barad's advice to look at '*phenomena - the ontological inseparability of agentially intra-acting "components"*' (Barad 2003: 815 italics in original). Intra-action is thus at the heart of a re-evaluation of agency, which means that the researcher is part and parcel of the co-construction of the phenomena, also via intra-action (Barad 2007: 139). The traditional separation of the observer and the observed, the separation of who knows what and what is known, becomes untenable (ibid.). This also raises questions about our relationships to the research participants, be they human or non-human (Iedema 2007), and creates an ethical position of absolute insistency on the fundamental responsibility we have as knowledge producers of the world and its becoming otherwise (Barad 2007: 183).

To advance a processual perspective of the helix model, and better grasp how it emerges and evolves, we will in the next section demonstrate the re-positioning of agency needed when technology, including the innovation to be commercialized, receives its own sphere. A focus on intra-actions makes it possible to include 'technology' as a new sphere, with possibilities to follow how the boundaries between the various actors are continuously re-constructed. To further explore this theoretical advancement of the helix model in practice, we will use two main empirical examples drawn from a broader qualitative study consisting of interviews, participant observations, and shadowing in the case of a regional development project in Uppsala. The two examples consist of an initial meeting between the regional office and the university, as well as an information meeting held with the general public. The project under study, a renovation of a bridge and inclusion of renewable technology to operate it, unfolded as a collaboration process around the quadruple helix model, involving researchers and academic entrepreneurs at Uppsala University, a small start-up, construction companies and consultants, Uppsala regional office and civil society. Throughout these two examples, we will zoom in on how the conceptual shift, from inter-actions to intra-actions, affects the possible conceptions we have of the dynamism involved in innovation system collaborations, suggesting that intra-actions will give both a deeper understanding of the difficulties at play in policy implementations, as well as explaining how artifacts, such as the technology, expose humans to uncontrollable events.

Empirical Example of Intra-Actions via the Fifth Sphere

The first technology that acts on the other more human-centered spheres is the ageing bridge. Before any implementation of the helix innovation system, the artifact thus enters the scene and makes the regional office aware of its responsibility to act on behalf of the Uppsala citizenry, facilitating their agency and need for commuting and recreation in nature. Later on, however, during the planning phase, the bridge is slowly accompanied by another technology. The renovation project takes a new direction due to how the logistical needs of the citizen are complemented by the need to respond to climate change. At this point, the human-produced scenarios of future climate changes have a greater effect on the renovation project in comparison to climate changes experienced in the form of flooding or storms. The scenarios translated into a broader environmental discourse (Darier 1999), intra-act with the regional office plans, whereby the renovation project takes a sudden turn. The personal concern of a public servant, and the notion of renewable energy technology as a solution to climate change, act in tandem on the possibilities to create a more sustainable and energy-neutral future. Renewable energy technology creates an opportunity, ready to be explored, in the renovation of the bridge.

Hence, before the first phone call and later meeting between the regional office and academic entrepreneurs, it was noticeable how the technology intra-acted, drawing these other spheres closer together. The technology re-constructs the boundaries (Holford 2013), as the regional office and university share a mutual devotion to an inclusion of renewable energy technology in the new updated bridge. The spontaneous mix of their roles evolves into something more seriously grounded during a face-to-face meeting between the regional office, university representatives, and one of the academic entrepreneurs. Mutual understandings emerge and are successfully aligned on common ground—renewable energy as a solution for a better future. The technical aspects of the technology enter the stage, as they are explained at length to the public servants with the help of a whiteboard. The budget is briefly broached, with considerations regarding the funding of the university research and possible development of the novel technology. The cost of implementation of the technology is balanced against the potential savings to be had when the bridge is electrified and its operation (lights, and opening and closing) is made self-sufficient. Technology intra-acts with the other spheres—regional office, university, and industry—thriving on ideas of how to make the specific bridge better, and how to make the region overall more entrepreneurial, sustainable, and competitive.

Without the opportunity created by the innovation in renewable energy technology at the university in the city, the renovation of the bridge would have been kept to exactly that, a renovation project. Now it has slowly transformed into a combined renovation, sustainability, and commercialization project. During this first meeting, something unfamiliar was thus made more familiar, and an attempt was made to reach a unitary whole in a common view on the promises of the technology (Shotter 2013). A sense of solidarity emerged, with intra-actions between the technology, the regional office, and the university, whose roles merged and thereby pushed the project forward in a new and still largely unknown direction.

With time, this new direction encounters the fourth sphere, civil society. At first, in the beginning of the planning phase, this sphere was only thought of as an actor who would later be included in the project. The strong intra-actions between renewable energy technology and the public servant in charge of the project also had an effect on the future possible connections between the technology and the general public. Civil society was imagined to want a more carbon-neutral Uppsala, much in accordance with the strategy and vision of the regional office. The technology, it was furthermore hypothesized by the regional office, would not only fulfil the future carbon-neutral goal of the city and county but could potentially draw in tourists. The renovation project quickly received an even stronger direction toward the inclusion of renewable energy technologies, as the idea to pimp the bridge with visible gadgets, such as lights and indications of renewable status, emerged. A portfolio of artifacts received agency, believed to open up for more human engagement with the bridge, increase environmental awareness, and perhaps even serve as an outlet for citizen concerns over climate change. However, this imagined public participation still had to be investigated in practice, by an actual inclusion of the fourth sphere. The intra-action with the general public was still minimal, played out as assumptions and ideas of their will to invest taxes in an updated recreational area around the bridge, complemented by the prospects of renewable energy technology. Consequently, the technology intra-acted on the deliberative democratic agenda, making it possible to connect around a seemingly more politically hot topic than just a 'renovation'. With this merger of ecological modernization and deliberative democracy, there were also more funding possibilities to approach, with the wish to extend the collaboration between the regional office and the university.

The collaboration developed over time, and the innovation system view grew stronger. Despite a funding application being rejected by VINNOVA (the Swedish state agency for innovation systems), the technology kept intra-acting in interesting ways, clarifying that it was no longer the researchers and academic entrepreneurs who were fully responsible for making the technology work. Rather, the river conditions, and hence 'nature', started to play an active role in the decisions made about the possibility to include an underwater current turbine in the design of the new bridge construction. Was there enough flow in the river to provide the required conditions? Nature and technology dynamically interplayed, offering themselves up as an engineering game of taming and control. At this point, the uncertainties grew among the public servants, while the university researchers and academic entrepreneurs faced business as usual: a professional task to be solved. The technology nonetheless intra-acted in a negative way. Instead of an opportunity for sustainability, it now posed an overly dynamic condition for the core of the project, the renovation.

Looking at the intra-actions among the grouping of spheres, there were boundaries created between various organizations involved in the bridge renovation. Geographical parameters played a significant role, as technology and humans at the university campus were excluded from intra-acting with the core technology and actors were devoted to the renovation only. More intense relations evolved between the regional office and other industrial actors, where human–non-human subjects became entangled in a more direct sense. The collaborative process took a direction where the sphere of the academic entrepreneurs and the sphere for the renewable energy technology were kept at a distance. This distance became more and more manifest over the following months, as the technology needed in-depth engineering research that took time. It was not properly included in the plans and project management tools for the renovation and was in addition playing a complementary role. It looked both unnecessary and unnecessarily complicated, even though everyone still agreed on its positive aspect with regard to the vision of a climate-neutral city and county.

Even though the regional office and industrial actors were grouped closer together, and the distance to the university and renewable energy technology grew, they still met briefly when the fourth sphere, civil society, was actively included in the innovation system. At an information meeting arranged by the regional office with the general public, we were not passive observers, but as knowledge producers we were entangled in the intra-actions that emerged, and according to our fieldnotes, the technology seemed to affect even us, as we encountered materialized elements of it during our participant observations:

> Most of the participants who attend live around the area; others are representatives from shipping lines and sailing clubs. The project leader is very active explaining and giving information about the renovation schedule and technicalities. People are interested in knowing about the noise during the construction phase, traffic issues, summer schedule for openings, and potential damage to their houses, but there is not much interest in the renewable energy technology. We therefore approached

> the project leader and asked him about the renewable technology explicitly. He explained and showed us that preparations were made for it in the drawings of the bridge as seen in Picture 1. A base plate for a 15.00-m turbine (Bottenplatta for Turbin 15.00 m) was specified in the architectural plan. He further explained that the talks between the local authorities and the professor were ongoing, and according to him, the negotiation was moving forward. (Fieldnotes, Informing the community, September 2015, also see García-Terán and Skoglund 2018)

Instead of focusing on the interactions between the government-university-civil society spheres, we were during this event surprised by the detailed drawings of the bridge, which actually included the renewable energy technology in a small way. The drawings and technology inclusion acted on us, as we continued to ask questions to better understand what for us was a change of direction of the renovation project. It was still noticeable, however, that the technology intra-acted more with us than with the fourth sphere, as the citizens mainly participated to inform themselves on the new bridge construction and updated recreational area. There was, therefore, neither resistance nor approval to the potential future renewable energy to be installed. The effort to accomplish a sustainable energy solution for the operation of the bridge was left blank. It seemed as though the drawings were there to intra-act with the university sphere rather than with the civil society sphere, and as the actual construction started, it was mainly the bridge that materialized, while a base foundation was added with possibilities to later add the renewable energy technology.

Discussion

The extension of the TH model and inclusion of the N:th number of spheres have previously been studied with a focus on interactions, for example, as developed between university-industry-government and civil society, with the aim to improve the collaboration between various actors (Leydesdorff 2010). However, due to a conventional focus on interactions between human agents and causal means-end processes, rather than emerging human–non-human relations and intra-actions, these studies have not been able to consider the addition of a non-human sphere in an ontologically intact and comprehensive way. As our detailed empirical examples of two meetings briefly illustrate, it is thus possible to expand the actors to be included in the grouping of spheres, and include a larger material arrangement, according to Barad's posthuman performativity (Barad 2003).

While a focus on interaction presumes the prior existence of independent entities, and a conventional understanding of mediation between them, our study has targeted 'emergence' by using the analytical tools of 'noticings', 'boundary re-construction', and 'intra-actions' (e.g., see Barad 2007: 139). In line with this, our empirical illustration sheds light on how the technology can be seen to intra-act, with more or less fruitful 'connections' developing dynamically between the different spheres in the implementation of a deliberative democratic agenda via the innovation system. Consequently, it is not the technology per se, as an additional sphere to the model, that should be seen as a 'unit' of analysis, neither is it the various human actors involved, but the locus for the analysis is the abstract coming together of these in the form of the intra-acts that unfold with material-discursive practices (Barad 2007: 141), and without any direct conclusions to be made about causation. It is rather fluid correlations that explain the direction of the project and what it is that drives forward the commercialization and materialization of an invention.

Hence, to study intra-actions is to grasp phenomena as inseparable/entangled, instead of dividing them between observer and observed (Barad 2007). Reality is composed of things-in phenomena, not of things-in-themselves or behind phenomena (ibid.: 140), where the fifth sphere (the technology) speaks and is spoken into 'becoming' via material-discursive practices. In comparison then, our focus on intra-actions that include the technology as an agent that is involved, in itself, exposes its limited capacity to affect the emergence of the commercialization if not fully taken on board and immersed in the other intra-actions, for example, those enacted via the general public. Not even if human proponents bring forth the promising green qualities with the technology, will the technology be able to 'disturb' a more linear approach to project implementation, from idea to materialization. It is clear that the technology would bring a green brand to the project, the bridge, and the region, but due to the fuzzy relations that it managed to shape, such a goal was with time abandoned.

Applying Barad's suggestion that we should shift focus from apparatuses as static prefab laboratory setups to rather try to understand the creation of phenomena via material-discursive practices, it also became clear how we as researchers influenced the phenomenon under study. By reflecting on intra-actions in relation to our own sensation of surprise at the inclusion of the base foundation in the bridge design at the information meeting, this made it clear that we were 'within' the observations, as an active part of creating the world (Barad 2007). Hence, participating at the information meeting made it possible to experience a deep sense of intra-actions, that is, the drawing as a human–non-human co-produced intra-actor that made us into specific intra-actors in our encounter with the project and its project manager. Hence, this is a form of analysis that can exemplify how we can get a more detailed conception of what it is that drives a commercialization process in a certain direction. Had a fifth sphere of technology not been added, this extra dimension would have been lost on us.

In comparison to how these intra-actions clarify how the renovation project emerged and changed direction infused by the technology, a conventional focus on interactions would thus have shown a more causal and human-led initiative. The technology would have been equal with the invention to be processed along the axis of the helix model. Looking at interactions, we would, for example, have observed how the regional office representatives in Uppsala mainly spoke about overarching goals of regional development according to an economic logic of new business establishment, an environmental logic with a set horizon of carbon neutrality, and a social logic that attempts to upgrade Uppsala to an attractive region for citizens—to grow its tax base and thereby possibility to develop the region according to the first two logics. And as a causal response, with a focus on interactions, we would have seen how the academic entrepreneurs aimed for a direct opportunity to develop a certain technology by prototype implementation—under the argument that relevant academic knowledge can only be generated, disseminated, and improved via test sites and experimentation. We would probably have observed more of their ambiguous application of project management tools, at the same time as their traditional research and theory contributions would be ongoing behind the scenes (Fowler et al. 2015). The researchers and academic entrepreneurs did, for example, express a strong wish to move from idea to materialization by receiving help to verify the 'real' qualities of the technology. In tandem with scientific contributions, this would facilitate commercialization and secure future scaling activities and investments in the technology. This latter logic for obvious reasons complied with the regional officers' wish to stimulate economic progress in the region, which is why interactions around this topic evolved more smoothly in comparison to the talk about research outputs and contributions to the academic community. Conclusively, in comparison to our interest in 'intra-actions', this focus on 'interactions' supports a traditional understanding of the innovation model, where each implementation of the QH model by some authority teases other actors to intervene in a preset macro perspective, for example, implemented by regional officers schooled in deliberative democracy and bottom-up empowerment. Indeed, the helix model does, just as the technology, function as an actor in itself, especially when it is implemented to accomplish public participation hand in hand with the facilitation of innovation for the transition of the energy system.

Conclusion

In knowledge production on innovation systems, scholars and policymakers often seek to optimize interactions between various actors that are linked in a causal way to the making of prosperous collaborations between them, assumed to result in successful commercialization processes and more sustainable futures. Even though these attempts have taken a processual view on how commercialization unfolds, this has nevertheless been pursued with a linear perspective in mind. Moreover, due to an intellectual foundation in various normative models that seek economic progression, for example, by the Triple and Quadruple

Helix models, policy-driven innovation systems have neglected interactions with non-human agents. Hence, with this idealism and bias toward human agency in the helix model for commercialization and citizen inclusion, we have explored the possibility to advance both the model as such and methods on how to study the implementation of the model.

Our study has resulted in two complementary shifts of the helix model. We have first added a fifth sphere with a non-human actor, in this case, a renewable energy technology acting as the innovation to be commercialized and materialized. Second, to be able to take this sphere into consideration as an equal agent in the grouping of spheres, we have made a conceptual shift from interactions to intra-actions. This development of the helix model affords both a less human-centered approach to agency, as requested in the shift to sustainability, as well as a more processual perspective on entrepreneurial endeavors. In addition, we argue that these shifts are paramount for a more transparent inclusion of the fourth sphere, civil society, and it is not until we properly admit the agency of a technology per se, that it is possible to effectively merge ecological modernization with public participation and democratic deliberation. A better understanding of how innovation models emerge hand in hand with citizen engagement is thus possible to attain via an analytical focus on 'noticings', 'intra-actions', and the re-construction of boundaries (Shotter 2006; Holford 2013; Barad 2003). With the help of these tools, it is thus possible to conceive innovation processes in more non-linear ways, as they dynamically evolve in the human–non-human nexus.

Hence, by outlining and testing a methodological approach for posthuman performativity, as suggested by Karen Barad (2003, 2007), we have been able to show how a fifth sphere can be taken into consideration as an agent in the implementation of an innovation system based on the Quadruple Helix model. Barad's approach is helpful, we argue, if the desire is to better understand a pluralism of posthuman relations, and how a particular technology can contribute to how innovative change unfolds in more or less 'successful' directions. As our two short empirical examples showed, the renewable energy technology that was supposed to be commercialized had less influence as an intra-actor in comparison to the bridge, which became more ingrained in the implementation. That is, by negating the static view of objects, we were able to affirm the uncertainty involved in commercialization processes, along with their dynamism and often haphazard outcome. Using Barad's understanding of apparatuses as material-discursive practices can thereby clarify how intra-acting objects transform other agents and the implementation of the innovation processes. What is important is thus the possibility to analytically see the agentic force of the technology, and the agency of the various spatial, temporal, and contextual components (Ringrose and Rawlings 2015). Hence, we argue that a conceptual shift, away from interactions toward intra-actions, may help policymakers to understand the evolving relationships between human and non-human agentic matter and the non-linear dynamism involved in innovation system collaborations.

Bibliography

Anshelm, Jonas, and Martin Hultman. 2014. *Discourses of Global Climate Change: Apocalyptic Framing and Political Antagonisms*. London: Routledge.

Barad, Karen. 2003. "Posthumanist Performativity: Toward an Understanding of How Matter Comes to Matter." *Signs* 28 (3): 801–831.

Barad, Karen. 2007. *Meeting the Universe Halfway, Quantum Physics and the Entanglement of Matter and Meaning*. Durham, NC and London: Duke University Press.

Brange, Lisa, Jessica Englund, and Patrick Lauenburg. 2016. "Prosumers in District Heating Networks–A Swedish Case Study." *Applied Energy* 164: 492–500.

Carayannis, Elias G., and David F. J. Campbell. 2009. "'Mode 3'and 'Quadruple Helix': Toward a 21st Century Fractal Innovation Ecosystem." *International Journal of Technology Management* 46 (3–4): 201–234.

Carayannis, Elias G., and David F. J. Campbell. 2010. "Triple Helix, Quadruple Helix and Quintuple Helix and How Do Knowledge, Innovation and the Environment Relate to Each Other?: A Proposed Framework for a Trans-disciplinary Analysis of Sustainable Development and Social Ecology." *International Journal of Social Ecology and Sustainable Development* 1 (1): 41–69.

Carayannis, Elias G., and David F. J. Campbell. 2011a. "Open Innovation Diplomacy and a 21st Century Fractal Research, Education and Innovation (FREIE) Ecosystem: Building on the Quadruple and Quintuple Helix Innovation Concepts and the 'Mode 3' Knowledge Production System." *Journal of the Knowledge Economy* 2 (3): 327.

Carayannis, Elias G., and David F. J. Campbell. 2011b. "Mode 3 Knowledge Production in Quadruple Helix Innovation Systems." In *Mode 3 Knowledge Production in Quadruple Helix Innovation Systems*, edited by Elias G. Carayannis and David F. J. Campbell, 1–63. New York, NY: Springer.

Carayannis, Elias G., and Ruslan Rakhmatullin. 2014. "The Quadruple/Quintuple Innovation Helixes and Smart Specialisation Strategies for Sustainable and Inclusive Growth in Europe and Beyond." *Journal of Knowledge Economy* 5 (2): 212–239.

Carayannis, Elias G., Thorsten D. Barth, and David F. J. Campbell. 2012. "The Quintuple Helix Innovation Model: Global Warming as a Challenge and Driver for Innovation." *Journal of Innovation and Entrepreneurship* 1 (1): 2.

Darier, Éric, ed. 1999. *Discourses of the Environment*. Malden, MA: Blackwell Publishers.

Erdős, Katalin, and Attila Varga. 2012. "The Academic Entrepreneur: Myth or Reality for Increased Regional Growth in Europe." In *Creative Knowledge Cities: Myths, Visions and Realities*, edited by Marina van Geenhuizen and Peter Nijkamp, 157–181. Cheltenham, UK and Northampton, MA, US: Edward Elgar.

Etzkowitz, Henry. 2003. "Innovation in Innovation: The Triple Helix of University-Industry-Government Relations." *Social Science Information* 42 (3): 293–337.

Etzkowitz, Henry, and Chunyan Zhou. 2017. *The Triple Helix: University–Industry–Government Innovation and Entrepreneurship*. London and New York: Routledge.

Felman, Shoshana. 2002. *The Scandal of the Speaking Body*. Stanford: Stanford University Press.

Fowler, Nina, Marcus Lindahl, and David Sköld. 2015. "The Projectification of University Research: A Study of Resistance and Accommodation of Project Management Tools & Techniques." *International Journal of Managing Projects in Business* 8 (1): 9–32.

García-Terán, Jessica, and Annika Skoglund. 2018. "A Processual Approach for the Quadruple Helix Model: The Case of a Regional Project in Uppsala." *Journal of the Knowledge Economy* 10 (3): 1–25.

Grundel, Ida, and Margareta Dahlström. 2016. "A Quadruple and Quintuple Helix Approach to Regional Innovation Systems in the Transformation to a Forestry-based Bioeconomy." *Journal of the Knowledge Economy* 7 (4): 963–983.

Gustavsson, Eva, Ingemar Elander, and Mats Lundmark. 2009. "Multilevel Governance, Networking Cities, and the Geography of Climate-Change Mitigation: Two Swedish Examples." *Environment and Planning C: Government and Policy* 27 (1): 59–74.

Helin, Jenny, Tor Hernes, Daniel Hjorth, and Robin Holt. 2015. "Process Is What Process Does." In *The Oxford Handbook of Process Philosophy and Organization Studies*, edited by Jenny Helin, Tor Hernes, Daniel Hjorth, and Robin Holt, 1–16. Oxford: Oxford University Press.

Hildingsson, Roger. 2010. "The Deliberative Turn in Swedish Sustainability Governance: Participation from Below or Governing from Above?" In *Environmental Politics and Deliberative Democracy, Examining the Promises of New Modes of Governance*, edited by Karin Bäckstrand, Jamil Khan, Annica Kronsell, and Eva Lövbrand, 145. Cheltenham, UK and Northampton, MA, US: Edward Elgar.

Hock Heng, Low, Nuratika Fatima Mohd Othman, Amran Mohammed Rasli, and Muhammad Jawad Iqbald. 2012. "Fourth Pillar in the Transformation of Production Economy to Knowledge Economy." *Procedia—Social and Behavioral Sciences* 40: 530–536.

Hoggett, Paul. 2011. "Climate Change and the Apocalyptic Imagination." *Psychoanalysis, Culture & Society* 16 (3): 261–275.

Höijer, Birgitta 2010. "Emotional Anchoring and Objectification in the Media Reporting on Climate Change." *Public Understanding of Science* 19 (6): 717–731.

Holford, David 2013. "Boundary (Re-)Constructions as Human-Nonhuman Intra-actions Within the Workplace." *International Journal of Business Anthropology* 4 (1): 15–37.

Hulme, Mike 2008. "The Conquering of Climate: Discourses of Fear and Their Dissolution." *The Geographical Journal* 174 (1): 5–16.

Iedema, Rick. 2007. "On the Multi-modality, Materiality and Contingency of Organizational Discourse." *Organization Studies* 28 (6): 931–946.

Langlais, Richard. 2009. "Swedish Municipal Responses to Climate Change." In *Planning for Climate Change: Strategies for Mitigation and Adaptation for Spatial Planners*, edited by Simin Davoudi, Jenny Crawford, and Abid Mehmood, 262. London: Routledge.

Latour, Bruno, and Steve Woolgar. 1986. *Laboratory Life: The Construction of Scientific Facts*. Princeton, NJ: Princeton University Press.

Lee, Young Hoon, and Young Jun Kim. 2016. "Analyzing Interaction in R&D Networks Using the Triple Helix Method: Evidence from Industrial R&D Programs in Korean Government." *Technological Forecasting and Social Change* 110: 93–105.

Leydesdorff, Loet. 2010. "The Triple Helix, Quadruple Helix, …, and an N-Tuple of Helices: Explanatory Models for Analyzing the Knowledge-Based Economy?" *Journal of Knowledge Economy* 3: 25–35.

Lindgren, Monica, and Johann Packendorff. 2002. "Interactive Entrepreneurship: On the Study of Innovative Social Processes." EURAM 2nd Annual Conference: Innovative Research in Management, 9–11 May 2002, Stockholm School of Entrepreneurship, Sweden.

Naturvetarna. 2019. "Jobba för Klimatet i Stället för att Strejka." https://www.naturvetarna.se/vi-erbjuder/tidning-och-nyheter/2019/agera-for-klimatet-i-stallet-for-att-strejka/.

Newell, Peter. 2020. "The Business of Rapid Transition." *Wiley Interdisciplinary Reviews* Climate Change: e670.

Nyberg, Daniel. 2009. "Computers, Customer Service Operatives and Cyborgs: Intra-actions in Call Centres." *Organization Studies* 30 (11): 1181–1199.

Öhrlund, Isak, Mårten Schultzberg, and Cajsa Bartusch. 2019. "Identifying and Estimating the Effects of a Mandatory Billing Demand Charge." *Applied Energy* 237: 885–895.

Packer, Jeremy, and Stephen B. Crofts Wiley, eds. 2012. *Communication Matters: Materialist Approaches to Media, Mobility and Networks*. New York: Routledge.

Peris-Ortiz, Marta, Jõao J. Ferreira, Luís Farinha, and Nuno O. Fernandes, eds. 2016. *Multiple Helix Ecosystems for Sustainable Competitiveness*. Cham: Springer.

Ringrose, Jessica, and Victoria Rawlings. 2015. "Posthuman Performativity, Gender and 'School Bullying': Exploring the Material-discursive Intra-actions of Skirts, Hair, Sluts, and Poofs." *Confero* 3 (2): 80–119.

Sarpong, David, Azley Abd Razak, Elizabeth Alexander, and Dirk Meissner. 2017. "Organizing Practices of University, Industry and Government That

Facilitate (or Impede) the Transition to a Hybrid Triple Helix Model of Innovation." *Technological Forecasting and Social Change* 123: 142–152.

Shotter, John. 2006. "Understanding Process From Within: An Argument for 'Withness'-Thinking." *Organization Studies* 27 (4): 585–602.

Shotter, John. 2013. "Reflections on Sociomateriality and Dialogicality in Organization Studies: from 'Inter-' to 'Intra-thinking'... in Performing Practices." In *How Matter Matters: Objects, Artifacts, and Materiality in Organization Studies*, 32–57. Oxford: Oxford University Press.

Steiber, Annika, and Sverker Alänge. 2013. "The Formation and Growth of Google: a Firm-Level Triple Helix Perspective." *Social Science Information* 52 (4): 575–604.

Steyaert, Chris 1997. "A Qualitative Methodology for Process Studies of Entrepreneurship: Creating Local Knowledge Through Stories." *International Studies of Management & Organization* 27 (3): 13–33.

Thunberg, Greta. 2019. "Sweden Is Not a Role Model," translated into English by We Don't Have Time.

Tsoukas, Haridimos, and Robert Chia. 2002. "On Organizational Becoming: Rethinking Organizational Change." *Organization Science* 13 (5): 567–582.

Uggla, Ylva. 2008. "Strategies to Create Risk Awareness and Legitimacy: The Swedish Climate Campaign." *Journal of Risk Research* 11 (6): 719–734.

Wittneben, B. F. Bettina, Chukwumerije Okereke, Subhabrata Bobby Banerjee, and David L. Levy. 2012. "Climate Change and the Emergence of New Organizational Landscapes." *Organization Studies* 33 (11): 1431–1450.

3

Failure to Act or Impossible Task? The Pursuit of Climate Justice and Energy Security Through Litigation

Judith Herbst and Deanna Grant-Smith

Introduction

Despite the gravity of climate change and the need for global action to reduce CO_2 emissions which has been recognized for well over a decade (Smith et al. 2009), there has been limited multilevel improvement worldwide. Storm surges and sea level rise threaten to displace residents from their homes, while livelihoods and public health may be negatively affected by extreme weather events which are increasing in intensity and frequency as a result of climate change. These impacts are not experienced equally and are compounded in many places by social and economic disadvantage, inadequate infrastructure, insufficient medical care, limited employment opportunities, and reduced access to clean water, energy, and sanitation (Dundas 2016).

J. Herbst (✉) · D. Grant-Smith
Queensland University of Technology, Brisbane, QLD, Australia
e-mail: judith.herbst@qut.edu.au

D. Grant-Smith
e-mail: deanna.grantsmith@qut.edu.au

E. V. Shabliy et al. (eds.), *Discourses on Sustainability*,
https://doi.org/10.1007/978-3-030-53121-8_3

Climate justice foregrounds the inherent ethical and political dimensions of climate change, positioning its management as a moral imperative (Aminzadeh 2006), and raising questions about the extent to which governments and other parties can, and indeed should be held accountable for negative impacts of practices and policies they continue to pursue. Often framed as competing policy interests (Bazilian et al. 2011), responding to climate change and ensuring energy security through the provision of "adequate, reliable supplies of energy at reasonable prices and in ways which do not jeopardize major national values and objectives" (Yergin 1988, 112) are two of the most pressing challenges facing governments across the globe. The impacts of climate change and unmet energy security are inextricably linked, with both problems resulting in potentially catastrophic consequences, especially for communities that have generally contributed the least to the impacts of climate change but are disproportionately affected the most (World Commission on Environment and Development 1987). Applying the lens of climate justice discourse exposes these imbalances by highlighting the inequities experienced by those whose needs are left unrecognized and ways in which they are excluded from environmental justice. By highlighting how individual basic needs and the ability to function in society are undermined (Dundas 2016), it can also explain citizen actions being undertaken to redress these injustices.

Growing recognition of "the urgency and uncertainty of the climate change agenda" and the fundamental rights of citizens "*in* climate change" (Steele and Gleeson 2009, 6) has spawned litigation focused on regulatory inaction (Burger and Gundlach 2017). This worldwide movement comprises children and other traditionally marginalized groups such as First Nations peoples, who in seeking resolutions are not only demanding respect, procedural justice, participation, and self-determination in decision-making (Schlosberg and Collins 2014), but they are also asserting their fundamental human rights to exist in a just society.

Drawing from the global growth in court cases against governments and fossil fuel companies for their failure to protect people from the impacts of climate change, this chapter considers legal challenges at multilevels (Burns and Osofsky 2009), and how these legal challenges might act as a catalyst to bring about adaptation and mitigation that "seeks to moderate or avoid harm or exploit beneficial opportunities" (IPCC 2014, 118) by rebalancing power supplies toward energy security and developing climate change policy which forces a global shift away from fossil fuel dependence toward the promotion of renewables.

Global and Legal Principles

A succession of international agreements established nonbinding carbon emissions reduction commitments and/or self-imposed targets to deal with climate change. Beginning in 1992, 154 industrialized countries signed an agreement at the United Nations Framework Convention on Climate Change (2010) to prevent dangerous warming from greenhouse gases, voluntarily lowering their levels to those of 1990 by 2000. After long deliberations, the Kyoto Protocol extended this framework, and became effective in 2005 for merely 34 countries and the EU, resolving to cut their greenhouse gas emissions to predominantly nationally determined measures or adopt a market-based mechanism for this purpose. Revisions to the Protocol expire in 2020, but its intent to "provide the architecture for the future international agreement on Climate Change" (United Nations Climate Change 2019b) was worthy.

The Paris Accord recently enacted a blueprint for all UN member states to build upon these platforms. 174 states and the EU ratified an agreement "by keeping a global temperature rise this century well below 2 degrees Celsius above pre-industrial levels and to pursue efforts to limit the temperature increase even further to 1.5 degrees Celsius" (United Nations Climate Change 2019a). The Sustainable Development Goals were consequently adopted by the United Nations (United Nations Department of Economic and Social Affairs n.d.), outlining 17 ambitious targets under its 2030 Agenda for Sustainable Development. Strategies for climate justice concern #7 Affordable and Clean Energy, and #13 Climate Action.

Another relevant international agreement, the Bali Principles of Climate Justice (International Climate Justice Network 2002), framed the issue of climate change within a human rights and environmental justice context. It was constructed by a coalition of groups that met in Johannesburg for the Earth Summit in 2002. This set of principles was intended to personalize climate change due to its life or death nature, particularly for those termed ecological refugees.

Attempting to Create Change Through Litigation

Due to the open-ended nature of these agreements and limited global policies for curbing greenhouse gas emissions, lawsuits have sparked across developed and developing nations to demand reform. Of the estimated 1000 cases of climate litigation, those integrating scientific data on the state of the climate with rights issues are raising opportunities to hold governments and polluters accountable for deleterious harm (Nachmany and Setzer 2018).

The decided court cases thus far have brought mixed results, especially those lawsuits that aim to stimulate "regulation through litigation." Nevertheless, policy failure is gradually being recognized with the issuance of a crop of favorable decisions. The public's appetite will be instrumental to keep mobilizing this movement although public failure to understand the risks of climate change coupled with self-interest over the long-term period in which these risks will materialize may undermine support for such policy change (Hersch and Viscusi 2007). Another possible impediment could be the uncertainty that some people feel about actual risks of climate change. In order for litigation to replace what is lacking in public policy, the courts have to offer incentives for controlling emissions at appropriate levels, to ensure they coincide with societal preferences, and importantly, to implement procedures for efficiency.

In addition, the chances for obtaining positive litigation outcomes may be explained by studies of economics literature where uncertainty or ambiguity to succeed has been documented. When confronted with ambiguous situations, people tend to take a risk-averse approach. They think it is better to have a 50 percent chance of winning a prize than a 50-50 option for something lesser known. Viscusi and Magat (1992) verified that this phenomenon, otherwise known as the Ellsberg Paradox (Ellsberg 1961), applies to environmental losses which are characteristic of climate change, but there is a caveat. When the probability of risk rises to an extreme level, then people tend to revert to a desire for uncertainty (Viscusi and Chesson 1999), possibly because they need to feel reassured that the risks might not be so great.

The growing body of evidence on the scientific, environmental, economic, social, and technological assessments of climate change impacts includes recommendations (IPCC 2014, 2015) that can dispel uncertainties, and foster rational decision-making over unease. It makes sense to invest today when emissions levels can be contained to prevent irreversible harms than to wait until they are irreversible (Arrow and Fisher 1974). Proponents advocate litigation can be a conduit to shape public policy and enhance performance by regulatory bodies, but critics argue that litigation is costly, lengthy, complex, and leads to unpredictable findings (Lytton 2007). Inconsistencies in findings are discussed in the following section on prominent climate change cases tested in courts.

A Landmark Ruling and Mounting Pressure of Civil Litigation

Historically, the majority of rulings in climate change cases ended in lower-court dismissals and reversals on appeals against plaintiffs. Breakthroughs are heralding glimpses of change as illustrated by the US where the highest number of climate justice cases have been tried (Setzer and Byrnes 2019). In *Mass v. EPA* (2007. 127 S. Ct. 1438), the United States Supreme Court delivered its landmark decision to uphold a favorable climate change result. Twelve states, four local governments, and multiple private organizations[1] petitioned against the Federal Government agency that is vested with statutory authority to regulate any substance that is deemed an air pollutant if there is evidence that it endangers the health or welfare of the public. By a slim majority, the Court agreed that carbon dioxide and other greenhouse gases met this standard to force the agency to regulate them under the *Clean Air Act* (1963. 42 U.S.C.

§ 7401). The justices deferred to environmentalists' tendency to employ regulation to prevent or rectify serious harm in the social and natural environment in spite of uncertainty or difficulty in measuring the effects from environmental disturbances (Cannon 2007).

This case thus represented a turning point in allowing access to the courts to exercise their opinions, giving liberal constructions of regulatory authority on an environmental matter. Petitioners successfully argued that systemic injury occurred. The majority opinion interpreted by Justice Stevens followed this chain of causation: motor vehicles emitted greenhouse gases which led to heightened greenhouse gases and sea level rise that reduced the Massachusetts coastline. Significantly, the Court was willing to infer injury resulted from disturbance to the climate system even though the effects were relatively small or not able to be readily measured as stated, thereby affording broad discretion by the Court.

Former President Bush subsequently implemented action by directing regulations to be imposed under the *Clean Air Act* (1963. 42 U.S.C. § 7401) to lower emissions from vehicles and improve fuel efficiency. Dissenters warned of the dangers of widening judicial restraint, arguing that regulatory matters should instead be prescribed by the legislative and then enforced by the executive branches of government (Cannon 2007). However, John Keegan, the US District Judge ruling in the initial case of *New York v. United States* (1992. 505 U.S.) warned that courts should not overstep their boundaries to prescribe policy as distinctive and independent roles are delegated to each branch of government. Most cases therefore enforced judicial restraint, abiding by narrow interpretations of regulatory authority.

In *Comer v. Murphy Oil USA* (Comer II) (2012. 839 F. Supp. 2d 849, 864–65 (S.D. Miss.)), the district court elaborated why such actions should be denied. The issues are fundamentally political, hence are not justiciable by virtue of the political question doctrine. The court referred to an earlier dismissal in *American Electric Power Company v. Connecticut* (2011. 131 S. Ct. 2527), which sought an injunction based on nuisance to require the utility to lower their greenhouse gas emissions. They held regardless of how well the case could be tried, neither the jurors nor judge have adequate resources and understanding that were entrusted to the EPA. Federal judges have neither economic, technological nor scientific resources which are available to a regulatory agency to deal with climate justice issues. When the courts have been receptive to consider such claims, however, experts have given testimony that is deemed sufficient for showing general causation which is difficult. To be able to assert a standing to sue or exercise judicial power under Article III, an ecological model must be applied. The legal test is showing an injury was real, particularized, imminent, or actual; it has to be causally related to the violation in question, and the relief that is requested must afford redress. However, to produce reliable evidence of specific causation relating to the defendants' alleged acts or omissions and the plaintiff's damages is improbable based on existing scientific reports (Peresich 2016).

A prima facie tort case is further complicated by the fact that duty, breach, causation and harm must be established by a plaintiff; in other words, the petitioner must show the defendant owed a duty of care, but the plaintiff was harmed from that breach (Goldberg and Zipursky 2001). However, it is problematic to link cumulative, indirect impacts of climate change to a particular activity of a polluting facility or government, and even harder to demonstrate a defendant made a significant contribution to its associated consequences (Peel 2011). Defendants will commonly justify that their emissions are minor in global terms. These types of debates in climate change litigation help to explain in part why the courts render different findings. Even if causation can be shown between an offending act and the inflicted injury, it raises other questions that require answers. For instance, if someone could trace emissions from a power plant to a resulting storm surge, what would be the duty of care that was owed by the operator? And if it was breached, to whom does the defendant owe a duty? A duty can only be owed to one or more victims not the world. So, which plaintiffs lie within the zone of foreseeable risk? And what would be an appropriate measure of liability considering multiple greenhouse gases interact during a period from exposure to harm (Hunter and Salzman 2007)?

In light of such challenges in proving common law allegations, mainly grounded in traditional common law allegations of negligence, actions over product defect and misrepresentation have been taken. Additionally, statutory claims have been raised involving nuisance and unfair trade practices which can all be categorized as tortious (Lytton 2007) due to their civil nature. Time and again, these cases have been unfruitful for petitioners.

Lawyers representing the state of Rhode Island confronted this obstacle by applying a different approach to attempt to demonstrate the merits of a lawsuit (*State of Rhode Island v. Shell Oil Products Company, LLC* et al. 18–395, D.C. R.I.). They are being joined by a dozen other U.S. municipalities in separately filed cases whereby plaintiffs contended climate change effects pertain to past behavior by fossil fuel companies that were aware since the 1970s that fossil fuel burning contributes to climate change (Hook 2019). They alleged 21 major oil and gas company defendants, including ExxonMobil and BP, knowingly marketed a harmful product while deliberately downplaying or denying its risk. For this reason, in spite of federal judges approving early motions to dismiss the complaints, communities and cities have filed appeals. They want the courts to assess the costs of climate-related impacts and attribute those impacts to the companies that can be linked to the generation of emissions (Hunter and Salman 2007).

Imposed threats of collecting evidence through the discovery process precipitated counterattacks by defendants. A targeted communications campaign and petition launched by ExxonMobil argued the climate change litigation is part of a "green conspiracy" (Schwartz 2016). Fossil fuel industries and governments are extremely powerful entities in these legal battles, and the courts may be unable to force offenders to address the global devastation that is forecast to exacerbate due to sovereign immunity and other doctrines that protect foreign states and individuals (UN Environment Programme 2017). However, there is a growing backlash by states alone fighting the Trump administration's decision to reverse executive orders undertaken to advance climate regulations overseas and domestically concerning the Clean Power Plan, vehicle fuel economy standards, methane leaks and HFCs (Adler 2019). And investigations are rising against companies for their failures to act. For instance, Exxon was brought up for charges of securities fraud relating to its failure to disclose its knowledge to its shareholders about its products leading to climate change impacts.

Researchers have made analogies between the current surge of climate change cases and earlier types of litigation for regulation in tort cases over various social issues to predict patterns of how the court rulings will unfold. For example, the literature draws comparisons with lawsuits against cigarette manufacturers to reduce smoking-related health problems and gun manufacturers to reduce violence which led to opposing legal outcomes (Lytton 2007). The outcomes at hand appear to be equally murky.

The Shift to a Human Rights Agenda

Advocacy groups and ordinary citizens are reframing climate justice issues as human rights violations which might offer better success. Human rights tribunals are the only available procedure to challenge governmental inaction for failing to protect victims' human rights. It lets global and regional complaints be invoked against nation states (where an individual is a citizen and a foreign state may be implicated in a transboundary action) by connecting environmental harm to international law (Posner 2007; Shelton 2010). Further, it overcomes restrictions, particularly by countries in the EU where there are more limited procedural rules—only allowing parties a standing to sue for challenges to the EU emissions trading scheme (Vanhala and Hilson 2013).

In human rights cases petitioners may argue, for example, that offenders infringed on victims' rights of being entitled to a healthy, productive life in co-existence with nature (Kiss 2010). Hence, plaintiffs are now not only suing governments for violations of human rights, but also are naming corporations in complicity and misleading shareholders about climate risks. Experts perceive these actions will not resolve the climate crisis, but they believe it could pose a significant difference to addressing it. And importantly, it could pressure governments to introduce more climate-friendly legislation, influence multinational corporations to lower their emissions, and widen public discourse (Osofsky 2010; Posner 2007).

As momentum is gaining for climate justice, these types of cases are springing up across countries, emphasizing a desire for cleaner energy to curtail emissions. A case from the Philippines by Greenpeace, individuals and groups[2] directly petitioned the state's human rights commission to consider the role of 50 of the largest fossil fuel companies (*Minors Oposa v. Secretary of the Department of Environmental & Natural Resources*. 1993, July 30. G.R. No. 101083, 33 ILM 173, at 187–8 (S.C.)), the "carbon majors" for violations of civic rights and duties under international law to prevent harm (Peel and Osofsky 2018). The Commissioner said the investigation would not be binding, however, he meaningfully drew on UN Guiding Principles on Business and Human Rights (OHCHR 2011). As a precursor to today's climate litigation, he suggested the findings would provide a forum for climate change dialogue based on credible scientific data.

Scientific evidence, in fact, prevailed upon the court in *Urgenda Foundation v. State of the Netherlands* (2015. Stichting Urgenda v. Government of the Netherlands (Ministry of Infrastructure and the Environment), ECLI: NL: RBDHA: 7145, Rechtbank Den Haag, C/09/456689/HA ZA 13–1396) when the Dutch nongovernmental organization and 900 citizens demanded governmental action to prevent climate change. In 2015, the Supreme Court of the Hague concurred with the defendants, and it ordered the Dutch government to reduce their emissions reduction targets by a minimum of 25 percent compared to 1990 standards to fulfill its duty of care to protect citizens from impending harm resulting from climate change. In making its decision, the Court accounted for the United Nations and European Union climate agreements coupled with the national civil code to determine the scope of duty of care, and they referred to causal evidence presented in IPCC Assessment Reports. They declared the government needed to take precautionary measures to mitigate damages as swiftly and extensively as possible to protect rights to life, home, and family life afforded by the European Court of Human Rights. Although the court handed down its judgment based on a tort cause of action, notably, the human rights arguments put forward by the nongovernmental organization, Urgenda, were used as an interpretative tool to show the government breached its duty of care (Peel and Osofsky 2018).

That same year, the Pakistan High Court went farther in handing down an historic decision in *Ashgar Leghari v. Federation of Pakistan* (2015, Sept. 4 and 14. Lahore High Court Green Bench, Orders). In light of Pakistan's strong vulnerability to the impacts of climate change, reflected in severe flooding in 2010 and 2011, the Justices agreed that the government's delay in implementing the nation's climate policy abridged the plaintiff's rights—an agriculturalist—to life, dignity of person, privacy of home and property under the 1973 Constitution (Articles 9, 14, 23 and 19A) (The Constitution of Pakistan 1973 and amendment 2010). The Court ordered the implementation of the country's Climate Change Policy (2014–2030) for adaptation as a priority by 2016 to safeguard the rights of a class of people including the poor or others who are vulnerable (Peel and Osofsky 2018). Moreover, the Court identified climate justice as a guiding principle. They highlighted that although the well-established foundation of environmental justice mostly surrounds localized issues, jurisprudence needs to gravitate to consider broader matters of climate change justice in recognition of its greater urgency and overpowering nature.

Greater Action by Marginalized Groups

Youth and First Nations people are heightening their legal campaigns. A remarkable outcome was seen in *Juliana v. United States* (2016, Nov. 10. No. 6:15-cv-01517, (D.Or.), 46 ELR 20175) where a youth-led coalition the plaintiffs argued the government is violating their substantive due process Constitutional rights to life, liberty, and property, and they challenged fossil fuel policy by demanding the government implement a recovery plan to lower emissions to mitigate effects grounded in climate science. They alleged a breach of public trust obligation happened by adults who are charged with responsibility to care and protect natural resources that are in the public interest (Peel and Osofsky 2018). This precedent-setting case focused on meeting the timeframe that scientists predict remains open to reverse the severe impacts of climate change. The inclusion of children in this and other cases disseminates powerful narratives through its anthropocentric emphasis, demonstrating its ability to evoke powerful responses within and outside the halls of justice (Rogers 2013). The lower court judge, in finding for the plaintiffs, determined a new right, that is, "the right to a climate system capable of sustaining human life…without which there would be neither civilization nor progress" (*Juliana v. United States*, 2016. Nov. 10. No. 6:15-cv-01517, (D.Or.), 46 ELR 20175, 32). This case was dismissed on appeal, but a request for review of the determination is expected. In another lawsuit, children in Alaska are bringing their case to the state Supreme Court to similarly adjudicate whether policy on fossil fuels is violating their constitutional rights to a safe climate (*Sinnok v. State of Alaska* (2019, Oct. 9. No. S-17297, (S.Ak)).

While these petitioners are busy fending off attacks to thwart their efforts, pioneering milestones have been achieved such as an appeal in Columbia when its Supreme Court ruled in favor of *25 Children and Youth v. Colombian Government* (2018. Supreme Court of Justice of Colombia STC4360-2018, 11001-22-03-000-2018-00319–01). Plaintiffs asked the government to act on deforestation and simultaneously protect their human rights. This decision was revolutionary, in that, the Court upheld the rights of future generations to a healthy environment and extended rights to the Amazon, a nonhuman entity, to be conserved. The Court also deemed that all affected parties should co-create an intergenerational pact for the Amazon for perpetuity through implementing updated land management plans to eliminate deforestation and mitigate emissions within only five months. This positive result portends that appeals for climate justice will be unrelenting until decisions are similarly favorable, especially in pockets that are hit hardest by impacts of climate change.

The consequences of climate change will be particularly devastating in low-lying coastal communities (Parry et al. 2009; Smith et al. 2009) where people are responding to a myriad of interconnected issues caused by sea level rise and extreme weather patterns such as diminishing land and declining access to freshwater and food (Adger and Barnett 2009). An illustrative example that brings together climate change and energy security can be seen in the actions of the First Nations people of the Torres Strait Islands, located at the northern tip of eastern Australia. Facing predicaments of losing their lives or claim to territories where they have resided for thousands of years, the Indigenous people filed a petition against the Australian Government for its failure to protect their homeland. This UN Human Rights Commission complaint alleges that the government is breaching their fundamental human rights under the *International Covenant on Civil and Political Rights* (ICCPR) (Australian Human Rights Commission n.d.) due to Australia's insufficient action on climate change. As a result, these communities are vulnerable to sea level rise, storm surge, coral bleaching, and ocean acidification (TSRA 2012). The major carbon contributor in the Torres Strait Islands is the electricity supply (TSIRC 2016) which feeds these isolated communities that are too remote to connect to the national power grid. As such, residents must rely on the local utility company that primarily generates electricity through diesel power stations (Ergon Energy 2019). The Islanders claim the government has infringed on their rights to life, culture, and freedom from arbitrary interference with privacy, family, and home, and is failing in both its climate change mitigation and adaptation responsibilities by having inadequate targets and plans in place to reduce greenhouse gas emissions, and by inadequately funding coastal defense and resilience measures. This petition is the first international level climate change investigation using violation of human rights as the underpinning argument, and the case is likely to have far-reaching implications for governments around the world.

Compensation for Successful Litigation

In previous litigation involving negligence complaints, judges determined compensation for breach of reasonable duty of care using risk utility analysis, multifactor balancing tests—accounting for a range of factors including convenience of administration, capacity of parties to bear the loss, prevention of future harm, moral blame (Dobbs et al. 1984), industry customs and overriding public concerns (Hunter and Salzman 2007). Scientific evidence continues to unequivocally identify profound effects of climate change on specific environmental systems, with risks predicted to grow more severe (IPCC 2014, 2015), so compensation for damage is thereby expected to increase. It might become easier to calculate estimates of at least the economic if not the intangible social costs from court assessments of damages to be granted in climate litigation. At the international level, however, there is no method for determining a traditional compensatory award. To invoke responsibility by a nation state, plaintiffs must instead rely on infringement of human rights, the no harm or precautionary principles and other applicable doctrines (Okowa 2006), and determine damages associated with loss of livelihood (Alabi 2012). Awards could be determined as an adaptation project and/or program.

For many cultures, the histories, identities—the fabric of its people's lives are linked to their homeland (Clifford 2001; Hau'ofa 1998), so intangibles may also need to be factored into awards. Further, courts may need to consider how to handle situations where territory becomes too perilous for people to stay in certain geographic areas. There is a tension in the strong physical and emotional attachment to place that pulls locals to remain where they dwelled for generations versus the push of an emergency that could cause people to have to leave (Smith 2013). For those who choose to remain rather than relocate to defend their territorial and human rights, it will be necessary to provide legal, political, cultural, and other support as conditions change (Farbotko et al. 2016).

At a minimum, remedies for defendants could amount to improving those natural resources that were damaged by their infringements such as implementing reforestation (Sigman 2007). A second approach could fall under the command and control style of regulation, to force an offender to either shift to lower carbon generation of electricity or find another way to increase fuel efficiency which could pose problems. The violator might move their operations to another jurisdiction where the constraints are less restrictive. Local action could thereby be undermined by leakage and diffusion of global greenhouse gases. Well-designed strategies could alternatively be trialed using new policies to stimulate industry toward broader regulation and adoption of energy-efficient technologies to curb emissions (Weiner 2007).

It should be recognized that legal enforcement might be inadequate because it functions as a one-off remedy. Greenhouse gases accumulate over time. Questions, therefore, remain over the efficiency and effectiveness of enforcing judgements. To be effective, compliance has to be monitored over the long term. This is why public policy is a better instrument than litigation because it can offer incentives. Pollution taxes and cap-and-trade pollution permits can not only be efficient but can also be more cost-effective than command-and-control schemes. Hence, litigation should only be utilized as a secondary response, and should be geared to the highest levels if public policy is impractical (Sigman 2007). Instilling coherent instructions though via public policy to counteract problems of climate change is the ideal way to resolve the dangerous impacts of climate change that may be incurred.

Since there is currently no international law to regulate global emissions and sources of greenhouse gases are widespread, the ability to achieve lower emissions is dependent upon cooperation by international actors who must consider the net benefits of formulating a treaty through mutual consent that has thus far escaped reach (Weiner 2007). These benefits may include protection of the climate, better public health, biodiversity conservation, enhanced reputation, and energy security. As witnessed in preceding cases, courts can adhere to state, national or international targets as guidelines for reducing greenhouse gas emissions to provide the basis for action notwithstanding its application to alleviate local ill effects (Peel 2011). Overcoming externalities of climate change can also be driven by heterogenous players taking advantage of lower renewable energy prices.

Today's competitive renewable energy marketplace can motivate the uptake of cleaner energy choices (IPCC 2018). Costs of wind and solar power have substantially decreased, even becoming cheaper than conventional fuels in certain markets (Cardwell 2014). Improvements in technology are bringing additional cleaner energy options with the advent of equipment to control pollution, capture and store carbon, and access devices to switch fuels. Vehicles can be converted to run on natural gas or biofuels (Baumert et al. 2005). And power can be generated from cleaner sources than fossil fuels. Australia, for example, has an abundance of wind and solar resources that can be harnessed to replace emissions from coal-powered stations (Yusaf et al. 2011).

Discussion and Conclusion

A key issue in making this transition is having sufficient capital to undertake any adaptation and mitigation initiatives (Williams and McDuie-Ra 2018). Presently, there is an implied lack of agency by residents in areas most affected by climate change in dealing with stakeholders who can provide the means to improve their conditions. If citizens or groups win court cases, however, addressing such complex challenges should happen through community engagement. Governments and companies can help plaintiffs by appropriating funds that were set up to accelerate carbon abatement to reshape better responses.

Communities will also require scientific assessments, planning help and resources to develop and implement adaptation strategies. Again, working in collaboration with stakeholders can help to build adaptive capacity of existing infrastructure, and to make shifts to renewable energy sources to mitigate extreme climate change consequences. Governments and industry should uphold human rights and enhance resilience of communities by adopting participatory forms of governance whereby communities can be partners in decision-making. It is recommended to spur community involvement in managing land and sea to preserve livelihoods, health and welfare, and to take a holistic approach that factors in local assets and skills. Often Indigenous communities have unique cultural resources, practical knowledge, and experience working with their neighbors to manage climate (Van Neervan 2015). This traditional ecological knowledge can be valuable to inform choices and develop appropriate responses to climate change (Dundas 2016).

At the moment, communities need to draw on their own knowledge of land, weather, cohesive family ties, and strength in self-organization. This storehouse of information is lacking in business planning or government strategy and policy decision-making, so policy consultation should be expanded to address the magnitude of the challenges currently being faced.

Signs of progress are being seen though with all signatories to the Paris Agreement having passed frameworks to address climate change. In addition, these laws and policies form part of more than 1500 endorsements for cleaner energy that include the establishment of funding bases to meet lower emission targets (Nachmany and Setzer 2018). Climate litigation may be important as a demand to fill the remaining gaps for action.

It is apparent that the scope of litigation for climate justice claims is being supplemented with new areas of jurisprudence. As litigation continues to develop legitimacy as a tool for climate justice and for influencing government policy and corporate behavior, the pursuit of human rights will likely factor more into judgments (Setzer and Byrnes 2019). Further, as permits for new emissions-intensive projects are being added to courtroom dockets, it could open up greater potential for compliance through legal obligations. Hence, the courts will not only consider the merits of the climate impacts to assert a standing to sue, but also will increasingly weigh the balance of public interest in potential benefits of a new project against its future harms (Peel and Osofsky 2018). Rulings thus far remain inconsistent, but a tide of change is likely to be ushered in. This is evidenced by a recent Australian ruling where the Chief Judge of the New South Wales Land and Environment for the first-time denied a proposal to start up a coal mining project in the Hunter Valley of New South Wales, Australia (*Gloucester Resources Limited v. Minister for Planning.* 2019. NSWLEC 7). Social impacts on locals that provoke a sense of "solastalgia," environmentally-induced stress caused by the loss of sense of place (Albrecht 2019), helped to sway the Judge in deciding this case.

When reparation is granted, different remedies may be available and desired by plaintiffs. The courts use different methods to calculate costs of avoiding injury versus damages caused by engaging in illicit behavior. A climate justice lens shows climate change is not only disproportionately affecting particular communities, but also how strategies in mitigation and adaptation can be driven by marginalized groups. Despite some court wins, it must be remembered that the ripple effects of proactive measures will only be felt in the distant future, if at all, without having mechanisms of global enforcement. Very little activity to reduce greenhouse gas emissions will alter global temperatures in the foreseeable future; and global warming caused by historical emissions are already stored in the oceans and atmosphere (Freeman and Konschnik 2014; Peresich 2016). Damaging impacts of natural disasters are forecast to continue which has significant implications for those seeking climate justice (IPCC 2015).

Nevertheless, cleaner energy systems are the most immediate, responsible, and forward path toward climate resilience and energy security. As the public becomes more vocal in exercising their rights to lower emissions and demand energy security through litigation and the uprising of activism and lobbying, it is possible for these interests to coalesce. A positive outcome would be to adopt meaningful energy policy and take other global responses to bring social change (Vanhala and Hilson 2013).

Notes

1. Petitioners comprised California, Connecticut, Illinois, Maine, Massachusetts, New Jersey, New Mexico, New York, Oregon, Rhode Island, Vermont, and Washington; District of Columbia, American Samoa, New York City, and Baltimore; and Center for Biological Diversity, Center for Food Safety, Conservation Law Foundation, Environmental Advocates, Environmental Defense, Friends of the Earth, Greenpeace, International Center for Technology Assessment, National Environmental Trust, Natural Resources Defense Council, Sierra Club, Union of Concerned Scientists, and U. S. Public Interest Research Group.
2. Greenpeace Southeast Asia and the Philippine Rural Reconstruction Movement petitioned the Commission on Human Rights of the Philippines on behalf of 13 organizations and 20 individuals.

Bibliography

Adger, W.N., and Barnett, J. 2009. Commentary: Four Reasons for Concern about Adaptation to Climate Change. *Environment and Planning A* 41, No. 12: 2800–2805.

Adler, D.P. 2019. *Litigation in the Year of Trump: Year Two.* New York: Sabin Center for Climate Change Law, Columbia Law School, June.

Alabi, S.A. 2012. Using Litigation to Enforce Climate Obligations Under Domestic and International Laws. *Carbon & Climate Law Review* 6, No. 3: 209–220.

Albrecht, G.A. 2019 *Earth Emotions: New Words for a New World.* Ithaca, NY: Cornell University Press.

Aminzadeh, S.C. 2006. A Moral Imperative: The Human Rights Implications of Climate Change. *Hastings International and Comparative Law Review* 30: 231.

Arrow, K.J., and Fisher, A.C. 1974. Environmental Preservation, Uncertainty, and Irreversibility. In *Classic Papers in Natural Resource Economics*, ed. Chennat Gopalakrishnan, 76–84. London: Palgrave Macmillan.

Australian Human Rights Commission. n.d. *International Covenant on Civil and Political Rights—Human Rights at Your Fingertips*. https://www.humanrights.gov.au/our-work/commission-general/international-covenant-civil-and-political-rights-human-rights-your (accessed November 3, 2019).

Baumert, K.A., Herzog, T., and Pershing, J. 2005. *Navigating the Numbers: Greenhouse Gas Data and International Climate Policy*. https://pdf.wri.org/navigating_numbers.pdf (accessed November 1, 2019).

Bazilian, M., Hobbs, B.F., Blyth, W., MacGill, I., and Howells, M. 2011 Interactions Between Energy Security and Climate Change: A Focus on Developing Countries. *Energy Policy* 39, No. 6: 3750–3756.

Bollen, J., Hers, S., and Van der Zwaan, B. 2010. An Integrated Assessment of Climate Change, Air Pollution, and Energy Security Policy. *Energy Policy* 38, No. 8: 4021–4030.

Brown, S.P., and Huntington, H.G. 2008. Energy Security and Climate Change Protection: Complementarity or Tradeoff? *Energy Policy* 36, No. 9: 3510–3513.

Burger, M., and Gundlach, J.M. 2017. *The Status of Climate Change Litigation: A Global Review.* United Nations Environment Program and Sabin Center for Climate Change Law: Columbia University. https://papers.ssrn.com/sol3/papers.cfm?abstract_id=3364568 (accessed November 1, 2019).

Burns, W.C.G., and Osofsky, H.M. eds. 2009. *Adjudicating Climate Change: State, National, and International Approaches.* New York: Cambridge University Press.

Cannon, J.C. 2007–2008. The Significance of Massachusetts v. EPA. *Virginia Law Review in Brief* 93: 53–62.

Cardwell, D. 2014. Solar and Wind Energy Start to Win on Price vs. Conventional Fuels. *New York Times*, November 23.

Clifford, J. 2001. Indigenous Articulations. *The Contemporary Pacific* 13, No. 2: 468–490.

Delmas, M.A., and Montes-Sancho, M.J. 2011. US State Policies for Renewable Energy: Context and Effectiveness. *Energy Policy* 39, No. 5: 2273–2288.

Dobbs, D.B., Keeton, R.E., Keeton, W.P., and Owen, D.G. 1984. *Prosser and Keeton on the Law of Torts* (5th Ed.). St. Paul, MN: West Publishing Company.

Dundas, A. 2016. *Towards Climate Justice: Decolonising Adaptation to Climate Change*. Uniting Justice. https://www.unitingjustice.org.au/environment/issues-papers/item/1216-towards-climate-justice (accessed November 15, 2019).

Duong, T.T. 2009. When Islands Drown: The Plight of Climate Change Refugees and Recourse to International Human Rights Law. *University of Pennsylvania Journal of International Law* 31: 1239–1274.

Ellsberg, D. 1961. Risk, Ambiguity, and the Savage Axioms. *The Quarterly Journal of Economics* 75, No. 4: 643–669.

Ergon Energy. 2019. Isolated and Remote Power Stations. https://www.ergon.com.au/network/network-management/network-infrastructure/isolated-and-remote-power-stations (accessed October 15, 2019).

Farbotko, C., Stratford, E., and Lazrus, H. 2016. Climate Migrants and New Identities? The Geopolitics of Embracing or Rejecting Mobility. *Social and Cultural Geography* 17, No. 4: 533–552.

Freeman, J., and Konschnik, K. 2014. US Climate Change Law and Policy: Possible Paths Forward. In *Global Climate Change and US Law* (2nd ed.), eds. Michael B. Gerrard and Jody Freeman, 1–67. Chicago: American Bar Association Section of Environment, Energy and Resources.

Goldberg, J.C., and Zipursky, B.C. 2001. The Restatement (Third) and the Place of Duty in Negligence Law. *Vanderbilt Law Review* 54: 657–750.

Hau'ofa, E. 1998. The Ocean in Us. *The Contemporary Pacific* 10, No. 2: 391–410.

Hersch, J., and Viscusi, W. 2007. Allocating Responsibility for the Failure of Global Warming Policies. *University of Pennsylvania Law Review* 155, No. 6: 1657–1694.

Hook, L. 2019. Oil Majors Gear Up for Wave of Climate Change Liability Lawsuits. *Financial Times*. June 9. https://www.ft.com/content/d5fbeae4-869c-11e9-97ea-05ac2431f453 (accessed November 5, 2019).

Hunter, D., and Salzman, J. 2007. Negligence in the Air: The Duty of Care in Climate Change Litigation. *University of Pennsylvania Law Review* 155, No. 6: 1741–1794.

Intergovernmental Panel on Climate Change. 2014. Annex II: Glossary (Katherine J. Mach, Serge Planton and Christoph von Stechow (eds.)). In *Climate Change 2014: Synthesis Report. Contribution of Working Groups I, II and III to the Fifth Assessment Report of the Intergovernmental Panel on Climate Change* [Core Writing Team, Rajendra.K. Pachauri and Leo A. Meyer (eds.)]. Geneva, Switzerland: IPCC.

Intergovernmental Panel on Climate Change. 2015. Annex II: Impacts, Adaptation, and Vulnerability, Vol. II: Regional Aspects, Working Group II Contribution to the Fifth Assessment Report 2014 (Vincente R. Barros and Christopher B. Field (eds.)). Arts 1, 6, 11 and 12. Cambridge, MA: Cambridge University Press.

Intergovernmental Panel on Climate Change. 2018. *Renewable Energy Sources and Climate Change Mitigation.* Geneva, Switzerland. https://www.ipcc.ch/site/assets/uploads/2018/03/SRREN_FD_SPM_final-1.pdf (accessed November 6, 2019).

International Climate Justice Network. 2002. *Bali Principles of Climate Justice*, August 28. https://www.ejnet.org/ej/bali.pdf (accessed November 6, 2019).

Kiss, A. 2010. The Right to the Conservation of the Environment. In *Linking Human Rights and the Environment*, eds. Romina Picolotti and Jorge Daniel Talliant, 31–46. Tucson: University of Arizona Press.

Lee, J., and Park, T. 2019. Impacts of the Regional Greenhouse Gas Initiative (RGGI) on Infant Mortality: A Quasi-Experimental Study in the USA, 2003–2014. *British Medical Journal Open* 9, No. 4: 1–7.

Lytton, T.M. 2007. Using Tort Litigation to Enhance Regulatory Policy Making: Evaluating Climate-Change Litigation of Lessons from Gun-Industry and Clergy-Sexual Abuse Lawsuits. *Texas Law Review* 86: 1837–1876.

Nachmany, M., and Setzer, J. 2018. *Global Trends in Climate Change Legislation and Litigation: 2018 Snapshot.* Grantham Research Institute on Climate Change and the Environment, April 30. https://www.lse.ac.uk/GranthamInstitute/?s=global+trends+in+climate+change+litigation (accessed November 4, 2019).

Nussbaum, M. 2003. Capabilities as Fundamental Entitlements: Sen and Social Justice. *Feminist Economics* 9, Nos. 2–3: 35–59.

OHCHR. 1966…in force 1976. International Covenant on Civil and Political Rights. New York, NY. December 16–March 23. https://www.ohchr.org/EN/ProfessionalInterest/Pages/CCPR.aspx (accessed November 3, 2019).

OHCHR. 2011. Guiding Principles on Business and Human Rights. https://www.ohchr.org/Documents/Publications/GuidingPrinciplesBusinessHR_EN.pdf (accessed November 3, 2019).

Okowa, P. 2006. The Legacy of Trail Smelter in the Field of Transboundary Air Pollution. In *Transboundary Harm in International Law: Lessons from the Trail Smelter Arbitration*, eds. Rebecca Bratspies and Russell Miller, 195–208. Cambridge: Cambridge University Press.

Osofsky, H.M. 2010. The Continuing Importance of Climate Change Litigation. *Climate Law* 1, No. 1: 3–29.

Parry, M., Lowe, J., and Hanson, C. 2009. Overshoot, Adapt and Recover *Nature* 458: 1102–1103.

Peel, J. 2011. Issues in Climate Change Litigation. *Carbon & Climate Law Review* 1: 15–24.

Peel, J., and Osofsky, H.M. 2018. A Rights Turn in Climate Change Litigation? *Transnational Environmental Law 7*, No. 1: 37–67.

Peresich, R.G. 2016. Summer. Climate Change Litigation. *The Brief; Chicago* 45, No. 4: 28–33.

Posner, E.A. 2007. Climate Change and International Human Rights Litigation: Critical Appraisal. *University of Pennsylvania Law Review* 155, No. 6: 1925–1946.

Rogers, N. 2013. Climate Change Litigation and the Awfulness of Lawfulness. *Alternative Law Journal* 38, No. 1: 20–24.

Schlosberg, D., and Collins, L.B. 2014. From Environmental to Climate Justice: Climate Change and the Discourse of Environmental Justice. *Wiley Interdisciplinary Reviews: Climate Change* 5, No. 3: 359–374.

Schwartz, J. 2016. Exxon Mobil Accuses the Rockefellers of a Climate Conspiracy. *New York Times*, November 21. https://www.nytimes.com/2016/11/21/science/exxon-mobil-rockefellers-climate-change.html?emc=edit_th_20161122&nl=todaysheadlines&nlid=3719225 (accessed November 6, 2019).

Setzer, J., and Byrnes, R. 2019. *Global Trends in Climate Change Litigation: 2019 Snapshot*. Grantham Research Institute on Climate Change and the Environment, July 4. https://www.lse.ac.uk/GranthamInstitute/?s=global+trends+in+climate+change+litigation (accessed November 4, 2019).

Shelton, D. 2010. The Environmental Jurisprudence of International Human Rights Tribunals. In *Linking Human Rights and the Environment*, eds. Romina Picolotti and Jorge Daniel Talliant, 1–30. Tucson: University of Arizona Press.

Sigman, H. 2007. Legal Liability as Climate Change Policy. *University of Pennsylvania Law Review* 155, No. 6: 1953–1959.

Sinn, H.-W. 2008. Public Policies Against Global Warming: A Supply Side Approach. *International Tax and Public Finance* 15, No. 4: 360–394.

Sinn, H.-W. 2012. *The Green Paradox: A Supply-Side Approach to Global Warming*. Cambridge, MA: MIT Press.

Smith, J.B., Schneider, S.H., Oppenheimer, M., Yohe, G.W., Hare, W., Mastrandrea, M.D., Patwardhan, A., Burton, I., Corfee-Morlot, J., Magadza, C.H.D., Füsser, H.-M., Pittock, A.B., Rahman, A., Suarez, A., and van Ypersele, J.-P. 2009. Assessing Dangerous Climate Change Through Update of the Intergovernmental Panel on Climate Change (IPCC) Reasons for Concern. *Proceedings of the National Academy of Sciences* 106: 4133–4137.

Smith, R. 2013. Should They Stay or Should They Go? A Discourse Analysis of Factors Influencing Relocation Decisions Among the Outer Islands of Tuvalu and Kiribati. *Journal of New Zealand and Pacific Studies* 1, No. 1: 23–39.

Steele, W., and Gleeson, B. 2009. *Planning in Climate Change: Toward a Relational Framework for Action.* Urban Research Program Research Paper 26. Brisbane, Griffith University.

Torres Strait Island Regional Council. 2016. *Our Carbon Footprint.* https://www.tsirc.qld.gov.au/changing-region/projects/our-carbon-footprint (accessed October 15, 2019).

Torres Strait Regional Authority. 2012. *Torres Strait Options to Reduce Regional Carbon Footprint.* CAT Projects and TSRA. https://www.tsra.gov.au/__data/assets/pdf_file/0005/2003/final20torres20strait20-20options20to20reduce20regional20carbon20footprint20-20january202012.pdf (accessed October 15, 2019).

UN Environment Programme. 2017. The Status of Climate Change Litigation: A Global Review. Nairobi, Kenya, May. https://wedocs.unep.org/bitstream/handle/20.500.11822/20767/climate-change-litigation.pdf?sequence=1&amp%3BisAllowed= (accessed November 6, 2019).

United Nations Climate Change. 2019a. The Paris Agreement. https://unfccc.int/process-and-meetings/the-paris-agreement/the-paris-agreement (accessed November 4, 2019).

United Nations Climate Change. 2019b. What Is the Kyoto Protocol? https://unfccc.int/kyoto_protocol (accessed November 4, 2019).

United Nations Department of Economic and Social Affairs. n.d. Sustainable Development Knowledge Platform. https://sustainabledevelopment.un.org/sdgs (accessed October 15, 2019).

United Nations Framework Convention on Climate Change. 2010. *Report of the Conference of the Parties on Its Sixteenth Session.* Bonn, Switzerland: UNFCCC.

United Nations General Assembly. 2009. Climate Change and Its Possible Security Implications. United Nations Security Council. https://www.securitycouncilreport.org/atf/cf/%7B65BFCF9B-6D27-4E9C-8CD3-CF6E4FF96FF9%7D/sg%20report%202009.pdf (accessed October 15, 2019).

Vanhala, L., and Hilson, C. 2013. Climate Change Litigation: Symposium Introduction. *Law and Policy* 35, No. 3: 141–149.

Van Neervan, E. 2015. The Country Is Like a Body. Right Now Essay Series. https://rightnow.org.au/topics/indigenous-people/the-country-is-like-a-body/ (accessed October 15, 2019).

Viscusi, W.K., and Chesson, H. 1999. Hopes and Fears: The Conflicting Effects of Risk Ambiguity. *Theory and Decision* 47, No. 2: 157–184.

Viscusi, W.K., and Magat, W.A. 1992. Bayesian Decisions with Ambiguous Belief Aversion. *Journal of Risk and Uncertainty* 5, No. 4: 371–387.

Wiener, J.B. 2007. Think Globally, Act Globally: The Limits of Local Climate Policies. *University of Pennsylvania Law Review* 155, No. 6: 1961–1980.

Williams, M., and McDuie-Ra, D. 2018. *Combatting Climate Change in the Pacific*. Cham, Switzerland: Palgrave Macmillan.

World Commission on Environment and Development. 1987. Report of the WCED: Our Common Future. New York: Oxford University Press. Retrieved from http://www.un-documents.net/wced-ocf.htm (accessed November 4, 2019).

Yergin, D. 1988. Energy Security in the 1990s. *Foreign Affairs* 67, No. 1: 110–132.

Yusaf, T., Goh, S., and Borserio, J.A. 2011. Potential of Renewable Energy Alternatives in Australia. *Renewable and Sustainable Energy Reviews* 15, No. 5: 2214–2221.

4

Perceptions and Awareness of Climate Change on Environmental Stewardship

Kimarie Engerman, Nisha Clavier, and Sharon Honore

Introduction

The debate surrounding climate change is ongoing.[1] Some believe that global warming is real and caused by human behavior. According to this school of thought, humans burn fossil fuels. Doing so results in loss of sea ice, stronger storms, global warming, and more droughts. Others believe doubt is raised by special interest groups funded by fossil fuel and related industries. This is because the gas emitted by humans is believed to be too small to cause changes to the earth's climate. The observed changes are possibly due to natural changes in the ocean current and the sun's heat. Yet, regardless of one's belief, the climate is changing and will

K. Engerman (✉) · N. Clavier · S. Honore
University of the Virgin Islands, Kingshill, USA
e-mail: kengerm@uvi.edu

N. Clavier
e-mail: nclavie@uvi.edu

S. Honore
e-mail: sharon.honore@uvi.edu

E. V. Shabliy et al. (eds.), *Discourses on Sustainability*,
https://doi.org/10.1007/978-3-030-53121-8_4

continue to do so in the future. In those circumstances in which climate changes cannot be avoided, humans should find ways to assist the environment. It is imperative, therefore, to understand the role humans can play in reducing the rapid rate of climate change.

Human reaction to climate change is based on social communication rather than immediate psychological change.[2] The visible impact of human activity and environmental change could take decades. For this reason, humans need to start engaging in environmental stewardship. Environmental stewardship was defined as "the responsible use (including conservation) of natural resources in a way that takes full and balanced account of the interests of society, future generations, and other species, as well as of private needs, and accepts significant answerability to society."[3]

There are various environmental stewardship behaviors that can be adopted. One behavior is growing fruits and vegetables. By planting, greenhouse gas carbon dioxide is removed from the air. This also reduces the need for fossil fuel vehicles that transport food to grocery stores. Another is being cognizant of energy needs and deciding to reduce electricity as much possible. A way to do so in the home is to use compact fluorescent light as opposed to incandescent light bulbs, and unplugging electrical devices when not in use. Moreover, individuals can walk or use a bike as opposed to driving a car. Additionally, recycling is another way to help the environment. It helps reduce trash. Water conservation is also helpful to the environment.

Engaging in the abovementioned behaviors can improve the quality of life around the world while maintaining a viable ecosystem and sustainable climate. However, improvement is all contingent on humans. Human perceptions and attitudes are necessary for environmental stewardship.[4] If humans are not aware of climate change issues and environmental stewardship behaviors they can employ, their behavior will continue to reflect a lack of care for the environment. Care of the environment is key for sustainability.[5] Furthermore, humans' perspectives can play an important role when examining environmental stewardship. Positive attitude, effort, and creativity were characteristics found in like-minded people working toward environmental stewardship.[6] Moreover, protection and stewardship of the environment is attributed to environmentally aware individuals; knowledgeable individuals tend to have a positive attitude toward the environment.[7]

Overall, changes in human behavior can help reduce climate change. However, humans must first be aware that there is a climate change issue and second, commit to environmental stewardship. Good environmental stewardship can reduce the misuse of environmental resources. This can somewhat help in the climate change issue. Given what is known about climate change, the purpose of this chapter is to examine the climate change awareness level of individuals on St. Croix, Virgin Islands. Climate change could possibly be attributed to the two Category 5 Hurricanes (Irma and Maria) that struck the United States Virgin Islands, two weeks apart in September 2017. The hurricanes caused major damage to the territories' infrastructure and way of life. The main source of the Virgin Islands' economy is the tourism industry. Climate change challenges the tourism industry.[8] Furthermore, climate change may create indirect environmental impacts such as water shortages, damages to landscapes aesthetics, and infrastructure. From a tourism standpoint in the Virgin Islands, this will include the inability to maintain local resort facilities, increased costs of maintaining infrastructure because of environmental changes, and extinction of valuable species due to changes in habitat. As a result of this, the long-term sustainability of the tourism industry will be in jeopardy. There will be a shorter season for tourism products, a limited number of natural sites to visit, and prices will increase to cover operating costs. The Virgin Islands cannot afford for this outcome to occur. Therefore, this study was justified.

The overall objectives of this research were to measure awareness of climate change and to assess whether there was a relationship between knowledge of climate change and environmental stewardship behaviors. In this research, the specific aim was to determine whether participants' residency status (Virgin Islands or non-Virgin Islands) was related to their knowledge of climate change and environmental stewardship.

Methods

The primary objective of PACCES was to design/test a methodological framework for assessing participants' awareness of climate impacts on their perceptions and attitudes of environmental stewardship. The case study involved residents on the island relative to geographic, racial, gender, age-related, and socioeconomic characteristics. The study was approved by the Institutional Review Board of the University of the Virgin Islands. Surveys were distributed to beachgoers at Rainbow Beach and Cramer's Park during weekday peak hours and on weekends. Researchers utilized a script when recruiting participants for the study. Part of the recruitment speech informed participants of the need for their consent, and their right to opt out of the study. Once verbal consent was received, participants were given a copy of the informed consent and asked to indicate their consent. Due to a lack of internet access, participants at Cramer's Park were given hard copies of the survey, while participants at Rainbow Beach utilized an iPad.

Participants

The study consisted of 352 individuals who were beachgoers on St. Croix, Virgin Islands. During the administration of the survey, participants were either attending Rainbow Beach (located on the west end of the island) or Cramer's Park (located on the east end of the island). 44.3% of participants were females and 55.7% were males. The racial background was 36.6% Black or African American, 39.3% Caucasian, and 14.9% Hispanic or Latino. See Table 4.1 for information on age categories and corresponding percentages.

Of the 319 participants who reported their residency status, the majority (69.8%) stated that they were residents of St. Croix. There was a reduction in the number of participants (292) willing to state which side of the island they resided on: 37.4% reported that they lived in Christiansted (east), 34.1% lived in Frederiksted (west), and 9.6% reported that their location was not applicable.

Table 4.1 Participant demographics—age

Category (Years)	Percent (%)
18–25	15.2
26–30	16.0
31–35	14.9
36–40	7.0
41–45	7.9
46–50	6.0
51–55	7.6
56–60	4.1
61–65	6.2
65 and older	6.8

Source Authors' calculation of participants' age

Instrument

The instrument used in the PACCES study is a combination of questions modified from the Climate Stewardship Survey[9] (Walker and McNeal 2013) and an article on 10 Ways to Be a Better Environmental Steward in 2018 (Kelly 2017).[10] The PACCES Survey consisted of six demographics, one cluster of 10 questions that measured environmental stewardship behaviors, and five clusters of 32 questions (total) that measured awareness of climate change questions. The measures were completed on a 4-point Likert scale (1 = definitely true and 4 = definitely false). The survey was administered in paper format, as well as electronically. An example of the questions asked can be found in Table 4.2.

Results

An analysis of the data showed that a participant's residency status significantly correlated with one of ten environmental stewardship behaviors: stop buying fashion that was produced quickly and inexpensively, usually at the cost of the environment. Forty-four percent of all residents reported that they stopped buying inexpensive fashions, in comparison to thirty-two percent of all nonresidents. The remaining environmental

Table 4.2 Cluster questions on PACCES survey

As an environment steward do you...	Which of these contribute to global warming...	A warming of the earth can cause...
reduce your meat consumption?	hurricanes?	disruptions in agriculture
change your mode of transportation?	volcanic eruptions?	changes in regional environments
educate yourself about environmental issues?	dust in the atmosphere?	an increase in the size of the ozone hole

Source Authors' survey questions, retrieved from the Perceptions and Awareness of Climate Change and Environmental Stewardship survey

stewardship questions yielded small differences among nonresidents and residents. For example, fifty-five percent of all nonresidents reported that they researched where their food came from, as did fifty-seven percent of all residents. See Table 4.3 for more information.

Participants' residency status was found to have significant relationships with their perceptions of events caused by the warming of the earth. A chi-square test of independence was performed on eight factors and six were found to have a significant relationship with residency status: disruptions in agriculture, changes in animal migration patterns, changes in regional environments, more UV radiation, sea level rise, and glaciers to melt. An increase in the size of the ozone hole and the dying of coral reef were the two perceptions that didn't yield significant relationships. Table 4.4 has more information on the significant values.

Residency also yielded six significant relationships with participants' perceptions of factors that contribute to global warming: hurricanes, automobiles/trucks, deforestation, burning fossil fuel for electricity, the hole in the ozone layer, and greenhouse gases. See Table 4.5 for an example of the percentages for participants' perception of hurricanes.

Residency had at least one significant relationship with each cluster of perceptions, except for participants' perception of whether the earth is cooling, or whether global warming is beneficial or harmful. Participants were asked to select the response, ranging from definitely true to definitely false, that best matched their perceptions. An example of one of the perceptions in the cluster was, the earth is cooling not warming.

Table 4.3 Participant environmental stewardship responses

As an environment steward, do you...	Yes (%)
Research where your food comes from and how it was produced?	
Resident	55.3
Nonresident	57.9
Reduce your meat consumption?	
Resident	56.1
Nonresident	46.8
Buy from local farms or start your own?	
Resident	75.5
Nonresident	76.6
Transform garden and other vegetable waste into a dark, rich, productive soil?	
Resident	38.2
Nonresident	36.6
Change your mode of transportation?	
Resident	29.5
Nonresident	29.3
Cut down on single-use plastics and microplastics or use reusable bags?	
Resident	89.2
Nonresident	90.3
Stop buying fashion that was produced quickly and inexpensively, usually at the cost of the environment?	
Resident	44.3
Nonresident	32.3
Educate yourself about environmental issues?	
Resident	81.9
Nonresident	86.2
Volunteer and working with others toward environmentally sound goals?	
Resident	59.7
Nonresident	51.1
Participate in public events, such as marches?	
Resident	52.0
Nonresident	42.0

Source Authors' calculation of age categories
Note The percentages of the two groups (resident and nonresident) are exclusive of one another

Table 4.4 Perceptions of events caused by warming of the earth

A warming of the earth can cause...	p =
Disruptions in agriculture	0.000
Changes in animal migration patterns	0.007
Changes in regional environments	0.020
More UV radiation	0.032
Sea level rise	0.005
Glaciers to melt	0.000

Source Authors' statistical analysis of a cluster of questions on the PACCES survey

Table 4.5 Which of these contribute to global warming?

	Hurricanes (%)			
	Definitely true	Probably true	Probably false	Definitely false
Resident				
Yes	49.8	34.4	13.0	2.8
No	27.2	40.2	23.9	8.7
	Automobiles/Trucks (%)			
	Definitely true	Probably true	Probably false	Definitely false
Resident				
Yes	43.1	31.5	17.1	8.3
No	76.9	18.7	2.2	2.2
	Dust in the Atmosphere (%)			
	Definitely true	Probably true	Probably false	Definitely false
Resident				
Yes	35.5	52.1	10.6	1.8
No	22.6	55.9	15.1	6.5
	Deforestation (cutting trees) (%)			
	Definitely true	Probably true	Probably false	Definitely false
Resident				
Yes	44.0	34.4	20.6	0.9
No	70.2	16.0	11.7	2.1
	An increase in the ozone hole (%)			
	Definitely true	Probably true	Probably false	Definitely false
Resident				
Yes	55.0	35.5	8.2	1.4
No	69.1	19.1	7.4	4.3

Source Authors' statistical frequencies based on responses received by participants on the PACCES survey

Discussion

This study is extremely important to the Virgin Islands, as there is a lack of emphasis on climate change and environmental stewardship. As of 2019, there were no recycling centers or programs on the islands. During the implementation of the study, many nonresident participants commented on the lack of recycling and the amount of trash around the island. Several expressed an interest in the study, and hoped that the results would trigger change in the territory. Additionally, there were residents who stated that they would support laws to create a more environmentally friendly island.

Moreover, it is generally felt on the island that individuals outside of the Virgin Islands (VI), mainly from the United States, have significantly more environmental stewardship behaviors than residents of the VI. However, these results indicate that there is not much difference between the reported environmental stewardship of residents and nonresidents. One confounding factor, which was not analyzed, was the number of residents who were not originally from the VI. This factor may attribute to the closeness in percentages of environmental stewardship.

The results of this study should encourage educational, community, and government agencies in the Virgin Islands to incorporate within their curriculum and/or programs, modules that teach citizens about climate change. Within this curriculum the importance, as well as the benefits of being environmental stewards, should also be taught. Creating programs geared toward youth will help in the fight to increase awareness of climate change and environmental stewardship.

As an outcome of this research, a brochure on the benefits of climate awareness and environmental stewardship was developed and distributed. In addition to the brochure, two proposals were created and submitted for funding to create junior clubs at local junior high schools. The first proposal, which received funding, created a stem exploration club at a junior high school. Approximately 30 students registered for the program and met weekly to participate in field trips and presentations from professionals within STEM fields. Students had an opportunity to go to a planetarium, the University of the Virgin Islands, and the airport to meet a pilot with a private plane. The program did well and was granted an extension for a second year.

The second proposal, which has not yet been funded, seeks to create the ACCES Junior Academy (AJA), an outreach component of the larger PACCES study. The purpose of AJA will be to educate students attending junior high schools on the importance of the climate and environment. AJA is also designed to teach students research skills through the implementation of a littering group project. AJA will not be a remedial club, so the focus will strictly be on learning about climate change, becoming an environmental steward, and completing the littering group project.

All in all, the results of the study will continue to be used to make a difference in the territory. Protecting the environment is everyone's responsibility. With the Virgin Islands being prone to hurricanes, it is critical that residents are fully informed about climate change. The overall quality of life for residents and tourism, the main source of economy, are severely impacted by hurricanes. Awareness can lead to more environmental stewardship by individuals and environmentally friendly practices by businesses, which can possibly help reduce future hurricanes. The findings of the program will be used to develop more programs for Virgin Islanders of all ages.

Acknowledgements This study is funded by the National Science Foundation, VI EPSCoR (MARE NOSTRUM CARIBBEAN: Stewardship through Strategic Research and Workforce Development) grant no. 1355437. The authors wish to acknowledge the significant contributions of Ms. Wanda Rosario and Mr. Matthew Perry.

Notes

1. Brian J.L Berry, Jayshree Bihari, and Euel Elliot. "The Limits of Knowledge and the Climate Change Debate." *Cato Journal*, 36, 3 (2016): 589–610.
2. M. Milenović, Snežana Živković, and Milan Veljković. "The Psychological Perspective of Climate Changes." *Casopis za Društvene Nauke*, 43, 3 (2019): 755–767. https://doi.org/10.22190/TEME190608046M.

3. Richard Worrell, and Michael C. Appleby. "Stewardship of Natural Resources: Definition, Ethical and Practical Aspects." *Journal of Agricultural and Environmental Ethics*, 12 (2000): 263–277. https://doi.org/10.1023/A:1009534214698.
4. Ajayi C., Omoogun, Etuki E. Egbonyi, and Usang N. Onnoghen. "From Environmental Awareness to Environmental Responsibility: Towards a Stewardship Curriculum." *Journal of Educational Issues*, 2, 2 (2016): 60–72.
5. Ron Sookram. "Environmental Attitudes and Environmental Stewardship: Implications for Sustainability." *The Journal of Values-Based Leadership*, 6, 2 (2013): Article 5.
6. Ann Carr. *Grass Roots and Green Tape: Principles and Practices of Environmental Stewardship* (Australia: Federation Press, 2001).
7. Norris I. Erhabor, and Juliet U. Don. "Impact of Environmental Education on the Knowledge and Attitude of Students Towards the Environment." *International Journal of Environmental and Science Education*, 11, 12 (2016): 5367–5375.
8. Don Webster. "Environmental Stewardship." In *Introduction to Tourism and Hospitality in BC*, edited by Morgan Westcott (Capilano University, 2015). Retrieved from https://opentextbc.ca/introtourism/chapter/chapter-10-environmental-stewardship/.
9. Scott, L. Walker, and Karen S. McNeal. "Development and Validation of an Instrument for Assessing Climate Change Knowledge and Perceptions: The Climate Stewardship Survey (CSS)." *International Electronic Journal of Environmental Education*, 3, 1 (2013): 57–73.
10. Carrie Kelly. (2017) 10 Ways to Be a Better Environmental Steward in 2018. Accessed March 7, 2018. https://www.ecowatch.com/environmental-stewardship-examples-2520328397.html.

Bibliography

Berry, Brian J.L., Jayshree Bihari, and Euel Elliot. 2016. "The Limits of Knowledge and the Climate Change Debate." *Cato Journal*, 36, no. 3 (Fall): 589–610.

Carr, Ann. 2001. *Grass Roots and Green Tape: Principles and Practices of Environmental Stewardship*. Australia: Federation Press.

Erhabor, Norris I., and Juliet U. Don. 2016. "Impact of Environmental Education on the Knowledge and Attitude of Students Towards the Environment." *International Journal of Environmental and Science Education*, 11, no. 12: 5367–5375.

Kelly, Carrie. 2017. 10 Ways to Be a Better Environmental Steward in 2018. Accessed March 7, 2018. https://www.ecowatch.com/environmental-stewardship-examples-2520328397.html.

Milenović, Miodrag, Snežana Živković, and Milan Veljković. 2019. "The Psychological Perspective of Climate Changes." *Casopis za Društvene Nauke*, 43 no. 3 (July): 755–767. https://doi.org/10.22190/TEME190608046M.

Omoogun, Ajayi C., Etuki E. Egbonyi, and Usang N. Onnoghen. 2016 "From Environmental Awareness to Environmental Responsibility: Towards a Stewardship Curriculum." *Journal of Educational Issues*, 2, no. 2: 60–72.

Sookram, Ron. 2013. "Environmental Attitudes and Environmental Stewardship: Implications for Sustainability." *The Journal of Values-Based Leadership*, 6, no. 2 (Summer/Fall): Article 5.

Walker, Scott, L., and Karen S. McNeal. 2013. "Development and Validation of an Instrument for Assessing Climate Change Knowledge and Perceptions: The Climate Stewardship Survey (CSS)." *International Electronic Journal of Environmental Education*, 3, no. 1: 57–73.

Webster, Don. 2015. "Environmental Stewardship." In *Introduction to Tourism and Hospitality in BC*, edited by Morgan Westcott. Capilano University. Retrieved from https://opentextbc.ca/introtourism/chapter/chapter-10-environmental-stewardship/.

Worrell, Richard, and Michael C. Appleby. 2000. "Stewardship of Natural Resources: Definition, Ethical and Practical Aspects." *Journal of Agricultural and Environmental Ethics*, 12 (January): 263–277 https://doi.org/10.1023/A:1009534214698.

5

Short-Sighted Visual Character Concerns in Renewable Energy Landscapes: A Case Study of South Australia

Brett Grimm and Joshua Zeunert

Introduction

The Intergovernmental Panel on Climate Change (IPCC) categorize electricity and heat production as the most significant category of human activity contributing to global greenhouse gas (GHG) emissions at 35%[1] (IPCC, 2015: 47, 88). Decarbonizing global electricity generation through transitioning to renewable forms of energy generation is therefore a central priority for climate change mitigation and improving sustainability. Despite the concept of shifting from finite fossil fuels to renewable energies being straightforward in theory, in practice it is a highly contested area, encompassing a multitude of socio-political, short-term economic and spatial complexities. In addition to existing production:

B. Grimm (✉)
Brett Grimm Landscape Architect (BGLA), Adelaide, SA, Australia

J. Zeunert
The University of New South Wales, Sydney, NSW, Australia
e-mail: j.zeunert@unsw.edu.au

E. V. Shabliy et al. (eds.), *Discourses on Sustainability*,
https://doi.org/10.1007/978-3-030-53121-8_5

> It will become increasingly difficult—if not impossible—to meet global energy needs without creating new energy landscapes. Such landscape shifts may be a difficult reality to accept, especially wherever people would prefer that landscapes remain unchanged indefinitely. (Pasqualetti and Stremke, 2018: 2)

To date in Australia, the involvement of built environment design professionals in the strategic planning of new *renewable energy landscapes* (herein REL) has been rare, especially as a targeted solution to GHG reduction. For the purpose of this chapter, REL constitute large-scale (>5 MW) wind and solar farms.[2] A noteworthy absence has been the discipline of landscape architecture, not only as the profession whose remit encompasses landscape scales and issues, but due to its prolonged voice in championing sustainability and Earth stewardship (McHarg, 1969; France, 2003; Weller, 2014; Zeunert, 2017; Fleming, 2019). Involvement in REL driven by an environmental sustainability agenda is lacking, as opposed to more commonplace interests that concern landscape character, visual impact and scenic amenity. Accordingly, helping to realize REL for their GHG mitigation and sustainability benefits should arguably be commonplace in landscape architecture.

Input into electricity production and distribution systems by landscape architects characteristically involves 'visual management' practices, centring on minimizing visual impact and protecting scenic amenity (Landscape Institute and Institute of Environmental Management and Assessment, 2013; Scottish Natural Heritage and the Countryside Agency, 2002; United States Department of Agriculture, 1995). Such visual landscape and 'visual management systems' are akin to concealment and even 'deception', 'exploiting the idea of nature to obscure the cultural reality of the landscape' (Wood, 1988). This practice has regularly occurred for both mine and resource-supply sites and fossil fuel power plants (and their associated and logistical infrastructures), electricity supply and distribution networks (primarily consisting of 'unsightly' high-voltage towers), forestry operations (Wood, 1988), desalination plants[3] and road systems (Appleyard, Lynch and Myer, 1964). A primary visual intent with these typologies is to maintain aesthetically pleasing landscapes, despite their often monolithic and scarring landscape characteristics. Yet as later discussed, achieving this can be at the expense of their greater impacts (such as GHG emissions) and environmental ramifications, analogous to placing 'parsley on the pig' (Zeunert, 2017: 125, 248).

To help realize increased and improved REL and move beyond existing shortcomings, limitations and constraints, landscape architects—spurred by a sustainability impetus—need to increasingly engage in shaping and evolving the strategic and statutory planning processes and legislative mechanisms that concern REL. This chapter seeks to contribute to this process, based on working within REL mechanisms in the State of South Australia (SA). We specifically examine the terrestrial component of REL—primarily wind and increasingly solar farms and energy parks. These spatially diffuse forms of REL often lead to conflict and protest in the communities where they are proposed and installed, and objections can be compounded due to their extensive spatial footprints and associated viewshed extents.

South Australia is globally noteworthy for its proportion of renewable energy generation, particularly its production through wind farms.[4] It achieves this against a national backdrop whereby 81% of Australia's electricity in 2018 was generated by non-renewables, primarily coal.[5] This chapter first examines SA's *energy generation mix,* specifically dissecting the implementation of wind energy generation to 2027. It then discusses the *planning processes* for wind farms in SA; the past two decades of statutory planning requirements are outlined, as well as current changes afoot relevant to wind and solar farms and emerging *energy parks.* Particulars of *landscape character* and *visual impact assessment* are outlined and illustrated. It is recommended that a statewide and consolidated landscape data set be developed that considers and integrates a greater spectrum of considerations in order to improve outcomes in the realization of emerging REL technologies.

South Australian Energy Generation

The past two decades have seen the composition of South Australia's electricity generation undergo substantial change. By 2018, all baseload coal-fired power stations were decommissioned in the State. Taking their place are numerous sizeable gas-fired turbines and renewable installations, which since 2018 have also been inclusive of utility large-scale battery storage.

In 2001 the Australian Government enacted the Renewable Energy (Electricity) Act 2000, otherwise known as the Mandatory Renewable Energy Target (MRET). This policy initiated the transitional phase to renewable energy and in South Australia helped lead to the implementation of wind farms. The MRET (2001) required large electricity retailers and consumers to gradually increase their output or purchase of renewable energy by 2%, from 10.5 to 12.5% by 2010. Across Australia, the MRET market penetration of renewable energy generation was met by 2007. In late 2007 there were 42 operating wind farms, equating to 563 turbines. At this time, South Australia possessed 51% of Australia's wind power generation, equating to 573 MW (with an additional 1986 MW in planning and feasibility phases) (Grimm, 2009). This is noteworthy considering SA's population was only 7.5% of Australia's total (ABS, 2008).

South Australia's total energy generation in 2018 was 13,506 GWh. Of this total electricity generation, 53% (7152 GWh) came from renewable sources, with 35% of total electricity generation coming from wind (Clean Energy Council, 2019). The total number of operational wind farms in SA was thirteen, with a total capacity of 1421 MW (AEMO, 2019a). South Australia is expected to have 73% renewable electricity by 2020–21, with a target of zero net emissions by 2050 (Australian Energy Market Operator, 2017) (Fig. 5.1).

South Australia's electricity makeup has been through major changes through a significant increase in inverter-based wind and rooftop Solar Photovoltaic (PV) capacity and a reduction in synchronous generation from coal and gas-fired sources (Fig. 5.2). Forecasted future energy

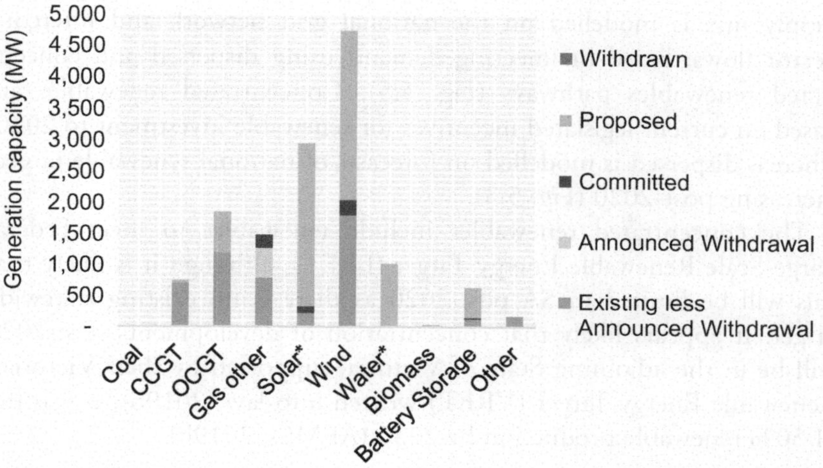

Fig. 5.1 South Australia energy generation mix in May 2019 (AEMO, 2019b)

Status	Coal	CCGT	OCGT	Gas other	Solar*	Wind	Water*	Biomass	Battery Storage	Other	Total
Existing	-	714	1,198	1,280	230	1,809	4	20	130	175	5,560
Announced Withdrawal	-	-	-	480	-	-	-	-	-	-	480
Existing less Announced Withdrawal	-	714	1,198	800	230	1,809	4	20	130	175	5,080
Committed	-	-	-	210	110	245	-	-	25	-	590
Proposed	-	45	659	-	2,620	2,694	1,025	15	488	-	7,545
Withdrawn	-	-	-	-	-	-	-	-	-	-	-

Fig. 5.2 South Australia Utility energy generation: current and proposed, in May 2019 (*Note* Existing includes Announced Withdrawal Solar excludes rooftop PV installations which are referred to as non-utility [AEMO, 2019b])

supply mix is modelled on the national grid network and interconnector flows, based on meeting demand using dispersed and concentrated renewables pathways (Fig. 5.5). Concentrated renewables are based on current legislated incentives for renewable investment to 2020, whereas dispersed is modelled on forecasts of national renewable targets increasing post-2020 (Fig. 5.3).

The concentrated renewables include renewables to meet Federal Large-Scale Renewable Energy Target (LRET), although it is likely that this will be limited in SA post-2020 as there is no existing statewide target. It appears likely that concentration of developments post-2020 will be in the adjoining State of Victoria, supported by their Victorian Renewable Energy Target (VRET) written into law (2019b), consisting of 50% renewable production by 2030 (AEMO, 2019b).

Wind Power Station	**Capacity (MW)**
Clements Gap	56.7
Hallett 4 North Brown Hill	132.3
Hallett 5 The Bluff	52.5
Hallett Stage 1 Brown Hill	94.5
Hallett Stage 2 Hallett Hill	71.4
Hornsdale Stage 1	102.4
Hornsdale Stage 2	102.4
Hornsdale Stage 3	112
Lake Bonney 2	159
Lake Bonney 3	39
Snowtown S2	270
Snowtown	98.7
Waterloo	130.8
Total	**1421**

Fig. 5.3 *Data* (*Source* South Australian operational Wind farms in May 2019 [AEMO, 2019b] https://www.aemo.com.au/energy-systems/electricity/national-electricity-market-nem/nem-forecasting-and-planning/forecasting-and-planning-data/generation-information)

Figure 5.4 illustrates forecasted energy generation by fuel source in South Australia over a 10-year period for concentrated renewables. It is noted that wind farms are predicted to increase by 1352 GWh by 2026–2027.

Dispersed renewables include the LRET supported investment, in addition to assumptions that further national renewable energy targets post-2020 will provide incentives.[6] The assumptions for modelling are a national target of 45% renewables by 2029–2030, which is justified as a mid-point between recent Queensland (50% by 2030) and Victorian State government announcements. In South Australia, it is predicted that reliance on baseload gas-powered generation will significantly decrease from 50% in 2016–2017 to 20–25% in 2026–2027.

Figure 5.4 illustrates the likelihood that South Australia will invest in additional renewable generation with wind, which is predicted to expand by 2583 GWh by 2026–2027. Additionally, solar is forecast to increase to 1617 MWh by 2026–2027. If further national renewable energy

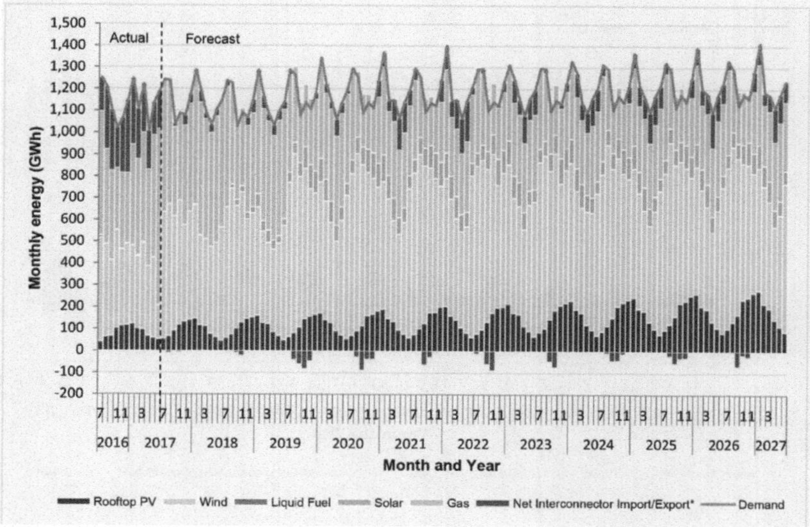

Fig. 5.4 South Australia forecasted *concentrated* renewable energy mix Australian Energy Market Operator (2017: 9)

targets are set beyond 2020, the South Australian generation market is predicted to become a net exporter to the eastern states from 2022.

Interpretation of Figs. 5.4–5.5 and the current environment for renewables in SA suggests that without further national or state policy encouraging additional investment, the proportion of renewable-generated electricity in SA is likely to stagnate. If this is the case, it is unlikely that SA will reach its target of 100% renewable energy generation by 2030.[7] SA was previously forecast to meet this target early, however, a change in state government in 2018 has seen the renewable energy market shift to the aforementioned receptive contexts in Victoria and Queensland.[8] This could diminish the status and trajectory of SA as a globally noteworthy generator of renewable energy, as multiple other regions may surpass SA's achievements.

Due to technology improvements, government policy and lower costs of infrastructure, the (non-utility) market is expanding from prior solely wind-powered installations. Recently, renewable proposals are

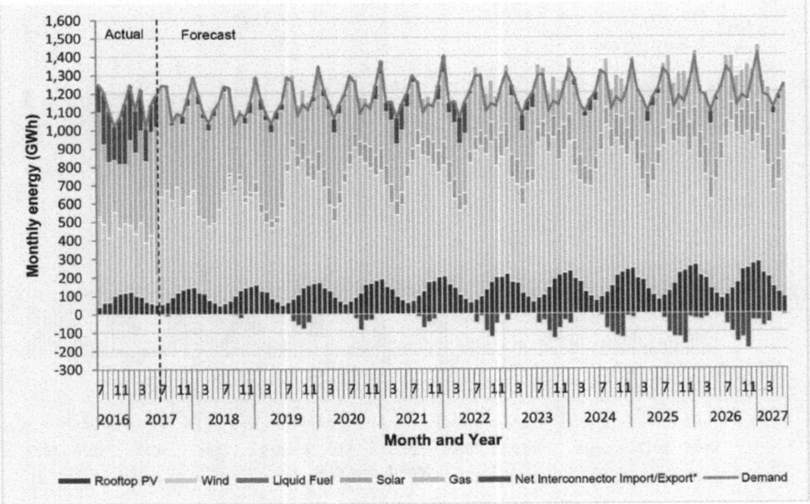

Fig. 5.5 South Australia forecasted *dispersed* renewable energy mix Australian Energy Market Operator (2017: 11)

comprising wind, solar and battery storage, which aim to provide consistent power supply for baseload. These developments are being referred to as 'energy parks' (DP Energy, 2019; Neoen Australia, 2019).

Of particular note in South Australia is the region of Port Augusta, where two coal power stations were recently decommissioned. The area is proposed to transition to a variety of renewable technologies including solar PV, solar thermal, wind, battery storage and pumped hydro, in the range of 1250 MW (approved but not constructed) with 800 MW storage, equating to a boost of ~65% in SA's renewable capacity (SA Planning Commission, 2019).

With regard to the planning and development of these diverse REL, it is imperative that landscapes are adequately evaluated for consideration of multiple qualities to ensure optimum land use outcomes, efficient energy generation and grid connectivity.

Planning Processes for Wind Farms in South Australia

A 'wind farm' is a group of wind turbines that are clustered together to generate electrical energy (Grimm, 2009). As wind turbines are optimally sited to maximize the capture of wind resources, such as on ridgelines, as well as spaced apart in order to not obstruct circulation of the wind direction, they present dominant visual elements in a given viewshed. Wind Farms are not a typical form of development. The collective agglomeration of significantly scaled vertical elements as per a wind farm had not been seen in Australia prior to the 1980s. Technological development has seen wind turbine heights increase substantially over the decades. In 1980, the maximum height to the tip of blade of onshore wind turbines was 50 metres (164 feet). In 2000–2005, this reached a maximum of 125–130 metres (410–427 feet). In 2019, the maximum height proposed in South Australia was 240 metres (787 feet) (Neoen Australia, 2019) (Fig. 5.6). This significant increase in height is attributed to the power generation per turbine, as output is relative to the diameter of the rotor. Since 2015, no new wind farm application in

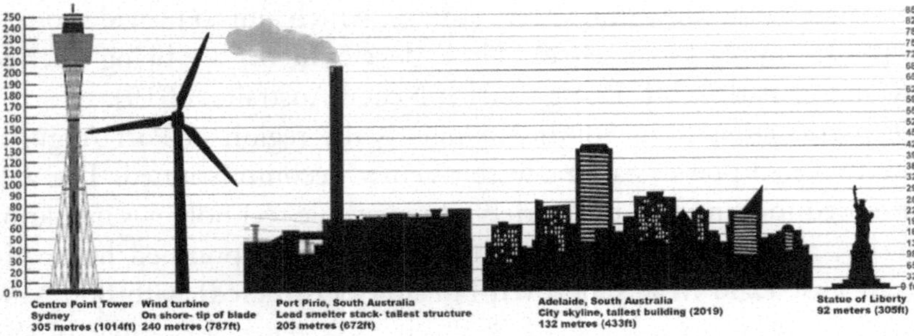

Fig. 5.6 Vertical scales of reference of a 240m turbine height to notable built elements (by authors)

South Australia has included a turbine with tip of blade height less than 150 metres (State Planning Commission, 2019).

The national uptake of wind farms in Australia has been sporadic. There is a multitude of factors that can enable or hinder installations, not limited to: cost competitiveness with coal and gas generators; age and life cycle of existing infrastructures; politics; network integration; environmental conditions; stability and baseload requirements; planning laws and community opposition.

In the early 2000s, a number of proposals for wind farms provided significant investment across South Australia and the adjoining State of Victoria due to ideal siting requirements, the availability of consistent wind resources, as well as proximity to transmission lines connecting population centres. Hall and Harvey (2006) note that the early adoption of wind farms in Victoria and South Australia took different approaches, 'Victoria has adopted a strategic approach which South Australia has a more reactive ad hoc approach' (Hall and Harvey, 2006: 721). In comparison to Victoria, it can be argued that SA's lack of red tape in planning restrictions and limited formal guidance in assessment methods supported early adoption of wind farms. It is noted more recently, however, that four wind farm project approvals (Stony Gap, Palmer, Mt. Bryan–Hallett 3 and Allendale) have been appealed and heard by SA's Environment Resource and Development Courts. The grounds of appeal have consistently referred to visual amenity, noise and proximity to dwellings.

The complexity of the wind farm infrastructure has challenged state-based planning systems and tools in South Australia. These mechanisms did not anticipate the scale and environmental characteristics of wind farms, and more specifically, for example, the dynamic movement and acoustic effects. Consequently, planning and assessment authorities (Councils, state commissions panels) and the general public have been challenged to adequately understand the context of the infrastructure requirements and consequently interpret and adapt the principles and objectives of development controls within the Development Act (1993). Difficulties have proven to be in applying practical and objective assessment criteria and in planning statements, which reference renewable energy in particular land use zones.

Early in 2002, the state government agency responsible for statutory planning released an Advisory Planning Notice for consultation on wind farms (Planning South Australia, 2002). The advisory notice identified the relevant authority and criteria to be considered in the assessment, and included acoustics, vegetation and habitat, aviation flight paths and landscape amenity. Within these varying criteria, landscape amenity was noted for its importance:

> Generally wind farms should avoid areas of scenic beauty/quality, particularly natural landscapes, main focal points including significant vistas and ridgelines, State and Local Heritage Place and areas of environmental conservation significance. (Planning South Australia, 2002)

The Advisory Notice was superseded by a Ministerial Plan Amendment Report (Planning South Australia, 2002). The report provides reference to the requirement for strategic policy to guide developers and the public in planning matters relating to wind farm proposals. As per the legislated Development Act in SA (1993), there are four different planning procedures employed to assess wind farm development proposals. Firstly, applications by the private sector; secondly, crown developments; thirdly, major developments and fourthly, projects involving electricity infrastructure.

Private developments are lodged with the relevant authority, typically the local Council or State Planning Commission Assessment Panel (SCAP) (formerly the Development Assessment Commission pre-2017). The authority assesses the application against the local Council's Development Plans, which designate cadastral boundary zones of land use character and development compliance criteria. The Development Plans provide objectives and principles for development control. Before 2012, wind farms assessed through this planning process have been considered category three assessment (on merit), due to non-compliance within relevant land use zones. Subsequently, local Councils—particularly those in regional and rural areas where developments have been proposed—have had difficulty in addressing applications due to limited reference criteria. This has placed notable pressure on the proponents of the wind farms in terms of extended timeframes and unknown investment risks. A number of proposals have been assessed and approved through this process, with a medium proportion subject to appeal and subsequent court hearings. The uncertainty in the application process and potential associated (legal) costs have provided a barrier to greater investment in wind farms in South Australia. The uncertainty of the process is not only felt by the proponent but also the regional communities where the wind farms are proposed. Wind farm proposals and developments in some of these communities have created bipolar community perceptions and thus divisive social conditions.

The conflict partly explains why many environmentalists support the concept of wind energy in the abstract (conservation) but may object to specific projects (preservation) in what has been called the NIMBY (Not in My Backyard) syndrome (Gipe, 1995: 257). The NIMBY effect has been a consistent undertone in community engagement and objection (Thayer and Hansen, 1989; Bell et al., 2005). To date, the landowners where the turbines are sited have received financial compensation for the lease of land in the order of $A10,000–15,000 per annum, per turbine. Such financial compensation is of significant direct return for primary producers, often battling fluctuating economic returns from agricultural activities due to seasonal weather variations and climate change. On the other hand, neighbouring properties receive no compensation, yet may suffer ongoing perceived impacts on their amenity. A number of proposals, specifically: Waterloo, Allendale, Palmer, Mt. Bryan and Stony Gap have all been subject to divisive community outcomes. Some wind farm developments provide a community fund to help support local infrastructure in regional townships to offset the imbalance of financial compensation (AGL Energy, 2019). This is not, however, a direct value received by those that feel they are more significantly impacted.

In 2012, a statewide Development Plan Amendment (DPA) was gazetted in SA, providing policies to support improved guidance on the integration of wind farms. A central feature of the policy is to generalize visual impact, which is one of the most contentious planning assessment concerns with wind farms in SA. This factor has been addressed by an unmistakable acknowledgement in the policies that wind farms may need to be sited (and therefore accepted) in visually prominent locations (Minister for Planning, 2012). The wind farm policy clearly supports and promotes wind farm development as an envisaged use in most rural areas, and one that in large part is designed to secure favourable outcomes for the proponent. The development controls require objectives to manage the visual effect, leaving ambiguous evaluations on what is deemed acceptable visual impact. Understandably, this is challenging for landscape visual impact assessment to quantify. Notable key features of the DPA are:

- Identification of wind farms as appropriate in general farming, primary production and rural zones to provide investment certainty for renewable energy development in South Australia;
- Clear setback distances from townships and dwellings to wind farms: currently 1 km from associated dwellings and 2 km from townships for settlement zones of urban areas;
- Supporting policies in respect to aviation safety, turbine design and operation;
- Clearer wording to reflect visual management of adverse effects;
- Public notification requirements that provide third parties with appeal rights against the decision of a planning authority only when set back distances are not met (Minister for Planning, 2012).

To support the Ministerial DPA (2012) the Local Government Association (LGA) developed a guide that summarizes the assessment criteria found in Development Plans. The LGA guidelines are intended to usefully identify and explain the key factors associated with the Development Plan policy to help planners determine whether particular wind farm proposals are an appropriate development or not. The items to be addressed are consistent with an Environmental Impact Assessment (EIA) and comprise (Table 5.1).

Table 5.1 Wind farm assessment criteria in SA Central Local Government Region of South Australia (2014)

Planning	• location and setting • surrounding land uses • nature of the built environment • the location of turbines and proximity to dwellings • local and arterial roads and network of proposed private access roads • impacts on key regional land uses (mineral exploration and extraction; • tourism • conservation parks and reserves • areas of significant landscape value • safety and operation of airports and landing strips • sea ports • cumulative impact
Technical	• wind farm lighting • geological landforms • geo-technical/ground stability • emergency and bushfire management • shadow flicker • reflection or blade glint • interference with television reception, radio, and other electronic equipment (i.e. GPS); • construction impacts, including site access routes for construction vehicles • site construction remediation

(continued)

Table 5.1 (continued)

Environmental	• character • landscape quality • visual significance and amenity of the area's natural environment • native vegetation • biodiversity and matters of national environment significance (EPBC Act 1999) • areas of heritage significance • sites of cultural significance and indigenous heritage • impact on surface and groundwater • soils and erosion potential • coastal erosion and/or inundation • noise
Management	• community and stakeholder engagement, nearby property owners • development staging and timing of each stage • construction costs, decommissioning and land rehabilitation • local economic benefits and employment generation

Crown developments, major developments and developments applied under Sect. 49A Electricity Act 1996 (SA) are generally projects supported by the State Government as public infrastructure. The authority for wind farm assessment is the SCAP. Upon the Planning Minister's declaration of project status, the SCAP will facilitate referral discussions with local Council and state agencies to determine the scope of project assessment requirements. The wind farm proponent is required to undertake studies to validate the impacts and benefits of the development with reference to relevant planning principles and objectives. The studies are submitted to the SCAP who provide an evaluation and recommendation to the Planning Minister. Under this process the Minister determines approval or refusal; following their determination there are no further rights of appeal by either the public or applicant. Consequently, the process typically provides more certainty for the proponent, albeit depending on the political cycle (Grimm, 2009).

As of 2018–2019, South Australia is undergoing the biggest reform of its planning system in 20 years. The renewed planning system is underpinned by the new Planning, Development and Infrastructure Act

2016. Within the reforms, the South Australian Planning Commission has most recently developed a Planning and Design Code discussion paper on renewable energy policy (SA Planning Commission, 2019). Over $3.3 billion worth of wind farm proposals (construction cost) were approved from 2012 to 2015 and $1.9 billion in 2019 (State Planning Commission, 2019). To ensure continued investment in renewable energy projects it is imperative that *the State Policies and Design Code* identify policy layers that reference renewable energy infrastructure.

The State Planning Policy '12: Energy', reinforces the importance of a consistent framework for evaluating landscape character and visual impact assessment. The policy states the following objective:

> Facilitate energy technologies that support a stable energy market and continued energy supply and do not adversely affect the amenity of regional communities. (SA Planning Commission, 2019: 63)

The policy content in the Code (Outback Areas) 'General Section—Infrastructure and Renewable Energy Facilities' is based on the current policy with enhancement from best practice results, operational project studies and lessons learnt over the past decade. Its proposed land use zones, in particular, the 'Rural Zone' and 'Remote Areas Zone', provide clarification on whether renewable energy facilities are encouraged or not. Furthermore, the definition of renewable energy is broadened to support innovations that integrate technologies such as battery storage, hydropower and *solar facilities*.

The renewed planning system will introduce overlays that specify restrictions on development due to environmental and or cultural sensitivities. Overlays contain policies and maps that show locations and extents of special land features or sensitivities, such as heritage places or areas of high bushfire risk, which can apply across one or more zones. Overlays are intended to be applied in conjunction with the relevant zone, however, where policy in a zone is in conflict with the policy in an overlay, the overlay policy will take precedence. Not dissimilar to the Victorian planning system, the proposed Significant Landscape Protection Overlay will identify significant landscapes where wind farms are discouraged.

Consequently, proposed wind farms will be 'restricted development' in areas of high landscape significance, designated in SA by:

- The Barossa and McLaren Vale Character Preservation Area Overlay (significant *wine* and tourism regions);
- The Significant Landscape Protection Area Overlay (the majority of which falls within the Flinders Ranges area, a significant conservation and tourism region);
- The Hills Face Zone (a significant metropolitan viewshed);
- A 'Peri Urban Zone' which generally applies through the Mount Lofty Ranges east of *the Greater Adelaide Area*, including the area around Mount Pleasant, and Mount Torrens.

Restricted developments can be refused approval without assessment, and approvals are subject to third party appeal rights. In all other areas, wind farms are 'performance assessed' developments that are to be assessed for consistency with the Code. Approvals are not subject to third party appeal rights unless classified as restricted (Finlaysons, 2019). Ultimately this will support REL development with limited opportunity for appeals. Draft changes to minimum planning policy requirements for wind farm proposals are shown in Table 5.2.

In summary, it is evident that South Australia's planning policies are being updated to keep pace with new and more efficient forms of energy infrastructure. The introduction of new state policies and the Planning and Design Code provides the opportunity to support the integration of renewable energy infrastructure and landscape amenity assessment overlays to ensure baseline aesthetics are appreciated and understood in pre-environmental appraisal. Further discussion on trends of technology and the need for strategic landscape character qualification can only further enhance the tools of the new planning system.

Table 5.2 Existing vs Proposed SA planning system policy requirements for wind farm proposals (SA Planning Commission, 2019)

Feature	Current System	New System
Policy	2 Kilometre setback from township and settlement zones and urban areas	2 km, plus 10 metres per additional metre over 150 metres in overall turbine height (measured from the base of the turbine) from township zones and the like
	1 Kilometre setback from non-associated dwellings	Setback 1.2 km from the base of the wind turbines
Referrals	Environment Protection Authority for advice	Environment Protection Authority for direction on: wind farms; energy recovery from waste; energy generation and storage over 30 MW
Planning Authority	Council Assessment Panels (for non-state sponsored projects)	Council Assessment Panels (for Performance Assessment)
	State Commission Assessment Panel (for state-sponsored projects)	State Commission Assessment Panel (for Restricted)
Role of Technical regulator	Certificate	Certificate
Assessment pathway	Merit	Performance Assessment on rural land (Rural Zone) Restricted in Significant Landscape Protection Overlay and Character Preservation Districts Overlay
Public notification	Category 2 where it meets the required setbacks	All wind farms will require public notification
	Category 2 in other cases	Additional notification and appeal rights for restricted

(continued)

Table 5.2 (continued)

Feature	Current System	New System
Environmental impact policy (vegetation removal and bird/bat strike)	General policy seeking minimization of impact	As per current system
Aircraft safety (visual and physical markers)	Referral to Civil Aviation Safety Authority (CASA) in areas of protected air space	As per current system
Noise	Environment Protection (Noise) policy requirements	As per current system

South Australian Planning Code and Other Forms of REL

As illustrated above, the proposed planning code provides improved guidance for wind farm and ultimately for other forms of REL proposals in terms of setback requirements and consideration of overlays. It is also recommended that solar farms should require more specific criteria for development control and assessment, as they are becoming larger and occupying increasing landscape expanses. Similar to wind farm development controls, utility-scaled solar farm proposals will require setbacks from conservation areas (500 metres); 100 metres setback from townships and rural living and 30 metres for all neighbouring property. Other notable controls suggested in the draft code are fencing typologies, habitat corridors and refuges and discouraging proposals of high environmental, scenic or cultural value (State Planning Commission, 2019).

With solar farms and unlike wind farms, planning consideration needs to be applied to the agricultural and soil productivity of landscapes, as solar farms limit the surface area available for cropping/horticulture and to a lesser extent, for grazing. Due to the nature of solar farms encompassing large tracts of land, exclusion from certain classes of agricultural land productivity needs to be considered and subsequently mapped as strategic agricultural landscape overlays. Like wind farms, environmental qualitative values of solar farms need to be considered. Similarly, landscape character and visual impact assessment are of relevance to the criteria of evaluation and consideration of planning overlays.

Landscape Character and Visual Impact Assessment

A number of guidelines and frameworks are referred to for landscape character and visual impact assessment in EIA (Grimm, 2009). The National Assessment Framework (2007) was developed to facilitate consistency in criteria and rigour of assessment methods in landscape character and visual impact assessment. The assessment process comprises two stages: preliminary landscape assessment and detailed visual impact assessment (Australian Wind Energy Association (Auswind) and Australian Council of National Trust, 2007). State-based guidelines have been developed in the states of Western Australia and Queensland (Western Australian Planning Commission Department of Planning and Infrastructure, 2007; Australian Institute of Landscape Architects [Queensland, 2018]). Landscape character assessment comprises site investigations, literature review of context and planning overlays and interpretation of the following landscape features:

- the topography of the land;
- the location, amount and type of vegetation;
- natural features such as waterways, coast, cliffs, escarpments, hills, gullies and valleys;
- visual boundaries between major landscape types;
- the type, pattern, built form, scale and character of development within the landscape, including roads, built form, tourist amenities and walking tracks.

Some landscape character assessment methodologies employed seek to quantify the character of the landscape through matrices, supporting the evaluation of visual change imposed by the proposed development (Grimm, 2009).

Visual Impact refers to the degree of visual change created by a proposed wind energy facility and depends on the extent of change to the landscape caused by the development. It takes into account:

- Zone of Theoretical Visual Influence (ZTVI);
- the overall visibility of the development;
- the locations and distances from which the development can be viewed;
- the significance of the landscape and proximity to sensitive areas;
- landscape values associated with nearby land, specified areas of landscape and environmental significance, specified coastal locations and areas identified to accommodate future urban growth;
- sensitivity of the landscape features to change.

The Grimke matrix is a model that has been used in a number of wind farm landscape character and visual impact assessments in South Australia. The method integrates geographic information systems (GIS) to spatially evaluate the landscape character and potential visual change. For a given proposal, areas of visibility are quantified in accordance to topographic screening and vegetation absorption from specified viewpoints (typically 6–8). Photomontages from viewpoints are then used to validate the visual effect. Scoring for identified viewpoints is provided through via an objective scoring matrix, facilitating the assessment of impact. Finally, the Grimke model generates a distance weighted interpolation map, presenting the likely visual effect of the proposed development (Grimm, 2009). The Landscape Visual Impact Assessment (LVIA) methodology is based on two assessment stages.

- The Landscape Character Assessment (LCA) is concerned with identifying and assessing the importance of landscape characteristics and the existing landscape quality;
- The LVIA aims to quantify the extent to which the proposed development is visible, as well as defining the degree of visual change to the defined landscape character.

The LCA and LVIA are used to draw a number of conclusions about the significance of the visual impacts of a proposed development to the surrounding site locality. The process of assessment and associated tasks are summarized in Fig. 5.7.

Figures 5.8 and 5.9 illustrate cartographic mapping in GIS used to evaluate the LVIA. Figure 5.8 provides reference to a Zone of Theoretical Visual Influence (ZTVI) of a proposed development, taking into account topographic form within the regional landscape to a 20-kilometre radius

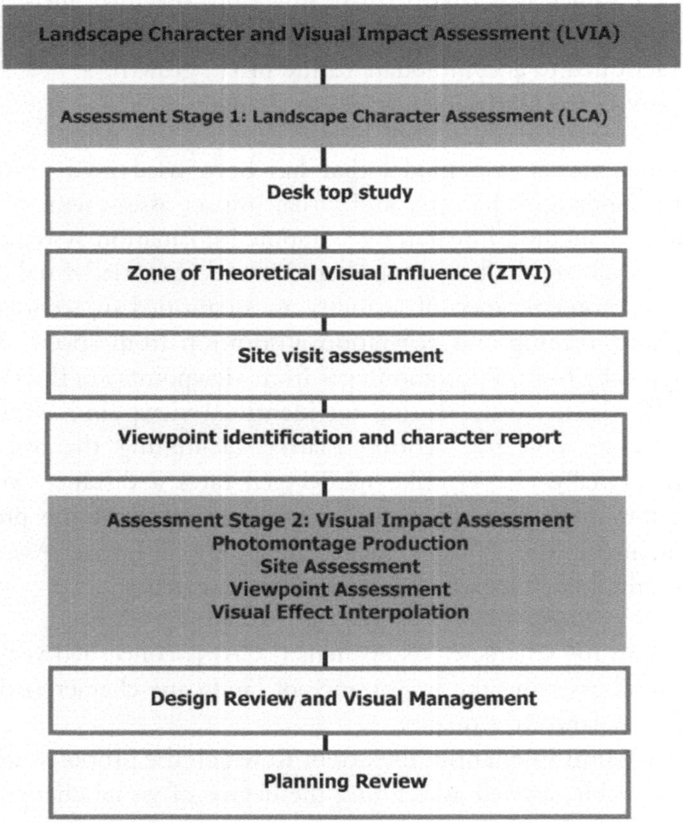

Fig. 5.7 LVIA—Two assessment stages and associated tasks GrimKe matrix (*Source* Wax Design and Brett Grimm Landscape Architect [2018])

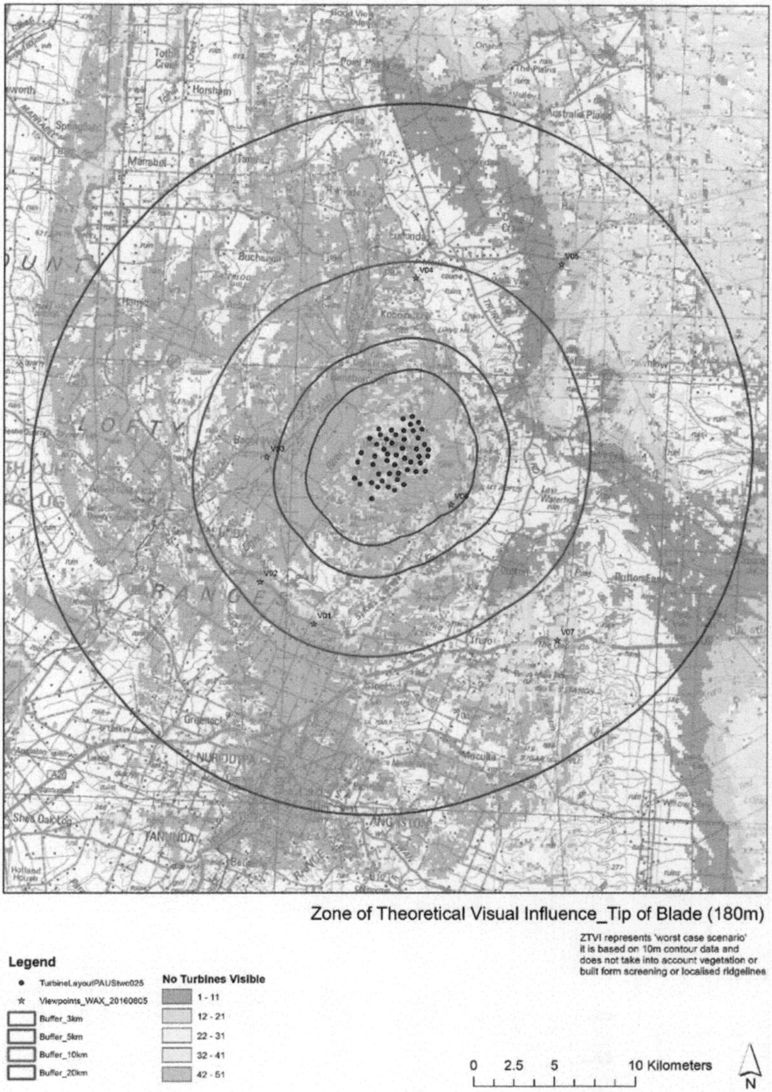

Fig. 5.8 Shows the various zones of theoretical visual influence in relation to the number and height of the turbines (*Source* Wax Design & Brett Grimm landscape Architect [2018])

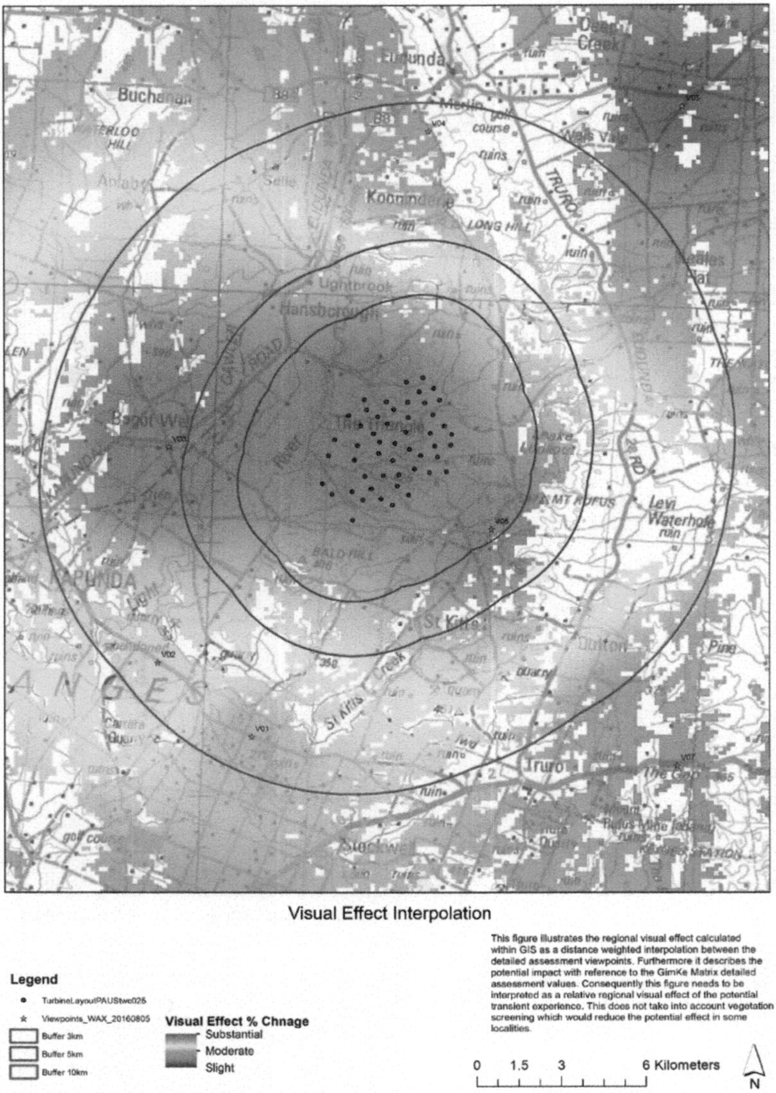

Fig. 5.9 Distance weighted visual effect evaluation of detailed viewpoint assessments (*Source* Wax Design & Brett Grimm landscape Architect [2018])

(12.4 mile). This is assessed onsite to evaluate and synthesize landscape character zones and typical viewpoints for photomontage production and detailed assessment. Figure 5.9 provides reference to the Visual Effect Interpolation which graphically represents the evaluation of visual change within the regional landscape as a distance weighted coefficient from the viewpoint detailed assessment using the GrimKe matrix (2018).

Prior studies have determined that the vertical nature of wind farms provides considerable view catchments, which can extend to 30–40 km depending on topography, vegetation screening and atmospheric conditions (Bishop, 2002; Sinclair, 2001; University of Newcastle, 2002). These studies were based on 70–100 m tip heights and accordingly, current potential view catchments are likely to be larger due to the aforementioned increases in wind turbine heights.

Generally in the South Australian context, most land uses surrounding wind farms remain unchanged. Grazing, for example, is the dominant landscape activity where wind farms are proposed. This occurrence provides the argument that the underlying land use remains unchanged, despite the addition of the wind farm. Conversely, solar farms are low-lying in relation to the ground plane, occupying a comparatively larger surface area, and thus, potentially reduce the capacity of the pre-existing grazing land use. Solar farms may also require the removal and/or regular pruning/grazing of vegetation above a necessary height (especially a shrub layer if present) as not to inhibit solar radiation reaching the PV panels. On the other hand, solar farms can be more readily visually screened and concealed by local topography (depending on conditions), creating less obtrusive viewsheds. Consequently, the context of REL evaluation and landscape capacity will require consideration of landscape character saturation as a trade-off for the efficiency of energy production per acreage developed.

Consideration of State-Wide Consolidated Landscape Data Set

The authors suggest that methods and approaches to improve REL planning, investment and implementation of wind farms and solar farms in South Australia could be conceptually modelled into a Geographic Information database to support the objectives of State Policy 12: Energy (SA Planning Commission, 2019). Currently in South Australia, data sets that cartographically depict values or classifications of land use and feasibility are piecemeal, which limits a strategic and statewide approach. It has been suggested that a data set—maintained by the State Government and open-source interface platform—could enable future clarity for identifying compatible sites to supply future energy demands and generate investment (South Australian Government Department of Premier and Cabinet, 2019).

McHarg's well-established 'sieve mapping' methodology can be utilized within GIS to concurrently analyse multiple spatial analysis overlays in order to synthesize and evaluate opportunities and constraints (McHarg, 1969). The digitization and analytic capacity of GIS provides an effective means to test and formalize capacity of scale, cumulative effects, efficiency and distribution proximity, land use and cultural sensitivity, terrain and accessibility. This process also facilitates a systemic assessment of criteria for determining suitability. A wind atlas has been developed providing clarity on average metre per second wind resource across the state, which can also be included as a data layer to support detailed studies (South Australian Government Department of Premier and Cabinet, 2019).

Consequently, there is a sound case for the centralized database to support commercial and state government partnerships in planning the future energy mix and regional landscape suitability, whether it be environmental preservation, primary production or settlement proximity. The planning systems' integration of modules and overlays provides an opportunity to support this data.

The detail of the planning overlays and guidance documents to inform EIA assessment methods is yet to be determined and would be dependent on SCAP and agency advice. Formalizing a landscape character and visual impact assessment framework would support consistency in the restricted applications, which are likely to be of notable due to the siting requirements of wind farms and vast areas required for large solar arrays or energy parks. Furthermore, cumulative effects of development proposals are challenging to evaluate, due to the lack of detailed data available (third party proponent intellectual property) and also the notion that each application is determined on its own merit. Consequently, a given proposal is verified by subjective interpretation of approved neighbouring developments and likely effects. With reference to the landscape character and visual effects, mapping a strategic overlay of landscape capacity could be ascertained as likely zones of theoretical visual influence, to provide baseline geographic information to inform investment planning and assessment.

Lothian (2000) developed a statewide landscape quality assessment that cartographically identified and valued the scenic quality of SA's landscapes. A digitized update of this study could support an overlay to reference areas of potential sensitivity to visual change. By referencing the likely statistical variance of wind farm perception within the various landscape quality zones, a map could be produced to illustrate areas of relevance for compatibility of development based on qualification of likely perceived sensitivity.

Community engagement of landscape values would qualify the amenity and scenic qualities in a strategic overlay. By orchestrating spatial quality assessments and compatibility of infrastructure, the current social planning issues of 'not in my backyard' could be verified in relevant data layers and referenced in the planning codes.

Conclusion

In an era now widely considered to be the 'Anthropocene Epoch' (Crutzen and Stoermer, 2000; Steffen et al., 2011), prioritization of scenic and visual concerns within the planning of energy landscapes appears very 'twentieth century' in the shadow of great extinctions (Wilson, 2016) and climate emergencies (Handmer et al., 2013; Leal-Filho, 2015). To improve global sustainability outcomes and reduce

GHG emissions, the transition to renewable energy landscapes should be a global priority. In the necessary shift to achieving more widespread acceptance of spatially diffuse REL, visual and scenic concerns are nonetheless an ongoing roadblock needing to be addressed in order to contribute to higher-level sustainability goals. Conceptually, a new aesthetic is required. Wolsink (2000) posits that there is a need to build *institutional capital*, whereby more open planning in REL can be fostered by developers and utility companies working more cohesively.

In order to be carefully considered, systematic and strategic across multiple and oft-conflicting concerns, REL need to integrate systems-thinking approaches. Planning assessments of REL could be based on a more strategic evaluation of efficiency and sustainability, supporting landscape capacity for compatibility with productivity. As a rare generalist and inclusive profession in an era characterized by disciplinary specialization and siloed approaches, landscape architecture should feature more prominently in the leadership and mix of professional consultants involved in REL, and expand upon the work in this vein (Grimm, 2009; Stremke and Dobbelsteen 2013; Sijmons et al., 2014; Zeunert, 2017: 156–165). As noted by Tudor (2014),

> Our landscapes are extremely important to us, they are part of our cultural heritage. With sympathetic planning, design and management they offer an opportunity to provide a more harmonious link between man and the natural world, for the benefit of both. Sensitive, informed, and integrated approaches should help us all to conserve, enhance, restore and regenerate landscapes that are attractive, diverse and publically valued, showing that environmental, social and economic benefits can go hand in hand. (Tudor, 2014)

Additionally, and somewhat ironically, landscape architecture's ongoing assistance in deceptive concealment (Wood, 1988) of spatially compact non-renewable electricity infrastructures is arguably unhelpful in increasing the uptake and acceptance of REL. Due to harnessing less energy dense renewable energy sources (wind currents, solar radiation) above the Earth's surface (van den Dobblesteen, 2014), REL are visually disadvantaged due to their comparatively expansive spatial footprints. In other words, visually concealing a wind farm that effectively harnesses wind 'resources' is a nigh on impossible task, especially with blades heights now equivalent to 70 story buildings. Due to their lower height, solar farms can be more effectively screened by landform, topography and vegetation, however, these can reduce the productivity of on-ground land uses. Nonetheless, surface land use might be considered secondary to the GHG mitigation benefits provided, depending on the given context.

It is worth noting that the existing uptake of REL has often been due to reduced financial risks presented to developers and investors; that is, within the current neoliberal free-market context, wind and solar farms have in many instances offered quicker and less risky financial returns than the alternatives. This is somewhat ironic, as neoliberal economics is routinely and rightly associated with planetary and social exploitation (Harvey, 2005). Similarly, and again ironically, developer-initiated projects (routinely lamented within rural (and urban) communities), and the shortcomings in South Australian planning processes facilitating these projects, have aided the uptake of renewable energy projects; albeit, in numerous cases, with piecemeal landscape and social outcomes. Despite reactive rather than proactive strategic planning instruments to holistically and clearly guide REL proponents and communities, significant GHG mitigation has nonetheless been achieved in South Australia's electricity market.

Finally, it should also be noted that the context discussed in this chapter—South Australia—presents conditions conducive to implementation of REL. SA has a low population density (1.78 people per km^2 (4.6/sq mi)); a low regional (versus urban) population; large spatial extents (983,482 km^2 (379,725 mi^2)); regions with ideal wind (and solar) resources; commonplace amenity landscapes (grazing and marginal cropping); low and marginal agricultural productivity (moderate and poor soils and low rainfall) and minor areas protected in conservation lands (in proximity to the national electricity grid). Despite these ideal physical and infrastructural conditions for REL, perhaps most noteworthy is South Australia's political realization of REL. Its significant proportion of wind and renewable energy generation has been achieved against a largely unreceptive federal governance backdrop, moreover characterized by climate change denial and continued support for and reliance upon fossil fuels in electricity production.

Notes

1. This is also true in Australia, the context of this chapter, being the largest at 33% in 2019, see http://www.environment.gov.au/system/files/resources/6686d48f-3f9c-448d-a1b7-7e410fe4f376/files/nggi-quarterly-update-mar-2019.pdf, page 9.
2. Wind and solar farms are being widely implemented across the globe and thus present existing and greater opportunity for involvement. Tidal energy is not currently widespread. Geothermal energy is also not widespread, and also does not present a spatially diffuse landscape footprint, which will be discussed later.
3. See https://www.aspect-studios.com/au/project/victorian-desalination-project-ecological-reserve/.
4. This statement is made in the context of renewable energy generation—such as wind and solar—in conventional contexts specifically seeking to mitigate greenhouse gases, as opposed to pre-existing contexts with hydropower (e.g. Tasmania) and geothermal (e.g. Iceland) renewable energy production.
5. See https://www.energy.gov.au/sites/default/files/2019_aes_table_o_march_2019.pdf, page 11.
6. See http://www.cleanenergyregulator.gov.au/RET/About-the-Renewable-Energy-Target/How-the-scheme-works/Large-scale-Renewable-Energy-Target.
7. See https://www.abc.net.au/news/2019-07-02/renewable-energy-targets-may-not-be-met-by-all-states/11267824.
8. In these States the proportion of energy generated by renewables is comparatively low, being 17% (Victoria) and 9% (Queensland) in 2018, see https://www.energy.gov.au/sites/default/files/2019_aes_table_o_march_2019.pdf, page 11.

Bibliography

ABS. (2008). *Population, Australian States and Territories*. Dec 2007. Accessed 26 October 2019. https://www.abs.gov.au/ausstats/abs@.nsf/mf/3239.0.55.001.

AEMO. (2019a, May 15). *Australian Energy Market Operator*. Retrieved from https://www.aemo.com.au.

AEMO. (2019b, May 15). *Key Connection Information Data File—Existing Generation*. Retrieved from https://www.aemo.com.au/energy-systems/electricity/national-electricity-market-nem/nem-forecasting-and-planning/forecasting-and-planning-data/generation-information.

AGL Energy. (2019, November 4). *Hallett Wind Farm*. Retrieved from https://www.agl.com.au/about-agl/how-we-source-energy/hallett-wind-farms.

Appleyard, D., Lynch, K., and Myer., J.R. (1964). *The View from the Road*. Cambridge: MIT. Press.

Australian Energy Market Operator. (2017, September 27). *South Australian Generation Forecasts*. Retrieved from https://reneweconomy.com.au/aemo-sees-south-australia-at-73-renewables-by-2020-21-21201/.

Australian Institute of Landscape Architects (Queensland). (2018). *Guidance note for Landscape and Visual Assessment*.

Australian Wind Energy Association (Auswind), & Australian Council of National Trust. (2007). *Wind Farms and Landscape Vlaues National Assessment Framework*.

Bell, D., et al. (2005). The 'Social Gap' in Wind Farm Siting Decisions: Explanations and Policy Responses. *Environmental Politics* 14 (4): 460–477.

Bishop, I. (2002). Determination of Thresholds of Visual Impact: The Case of Wind Turbines. *Environment and Planning B Planning and Design* 29: 707–718.

Central Local Government Region of South Australia. (2014). *Wind Farm Development Guidelines for Developers and Local Government Planners, Central Local Government Region of South Australia*.

Clean Energy Council. (2019, November 4). *Clean Energy Australia 2019 Annual Report*.

DP Energy. (2019, November 4). *Port Augusta Renewable Energy Park*. Retrieved from https://www.dpenergy.com/hybrid/port-augusta-renewable-energy-park/.

Finlaysons. (2019, October). *Proposed New Planning Rules for Wind Farms and Solar Farms in SA*.

Fleming, Billy. (2019*). Design and the New Green New Deal, Places*. Retrieved April (4 May, 2019), from https://placesjournal.org/article/design-and-the-green-new-deal/?cn-reloaded=1.

France, R. (2003). Green World, Gray Heart?: The Promise and the Reality of Landscape Architecture in Sustaining Nature. *Harvard Design Magazine* (18). Viewed December 15, 2015. http://www.harvarddesignmagazine.org/issues/18/green-world-gray-heart-the-promise-and-the-reality-of-landscape-architecture-in-sustaining-nature.

Gipe, P. (1995). *Wind Energy Comes of Age*. Chichester, UK: Wiley.

Grimm, B. (2009). *Quantifying The Visual Effects of Wind Farms: A Theoretical Process in an Evolving Australian Visual Landscape*. Dissertation for doctor of philosophy, University of Adelaide, Adelaide.

Handmer, J., Dovers, S., and Handmer, J. (2013). *Handbook of Disaster Policies and Institutions: Improving Emergency Management and Climate Change Adaptation*. Abingdon, Oxon: Routledge/Earthscan.

Hall, A., & Harvey, N. (2006). Environmental Assessment Procedures for Wind Power Projects in Australia. *Energy and Environment* 5 (17).

Harvey, D. (2005). *A Brief History of Neoliberalism*. Oxford: Oxford University Press.

IPCC. (2015). *Climate Change 2014: Synthesis Report*. Geneva, Switzerland: IPCC.

Landscape Institute, and Institute of Environmental Mananagement and Assessment. (2013). *The Guidance for Landscape and Visual Impact Assessment*. London: Taylor and Francis.

Landscape Institute, and the Institute of Environmental Management and Assessment. (2002). *Guidelines for Landscape and Visual Impact Assessment*. London: Taylor and Francis.

Leal-Filho, W. (2015). *Handbook of Climate Change Adaptation*. Heidelberg: Springer Reference.

Lothian, A. (2000). *Landscape Quality Assessment of South Australia*. Department of Geographical Studies, University of Adelaide, PhD.

McHarg, I. (1969). *Design with Nature*. Philadelphia, PA: Falcon Press.

Minister for Planning. (2012, October). Statewide Wind Farms Development Plan Amendment. Canberra, ACT: South Australian Government Gazette.

Neoen Australia. (2019a, November 1). *Crystal Brook Energy Park*. Retrieved from https://crystalbrookenergypark.com.au/.

Neoen Australia. (2019b, November 4). *Crystal Brook Energy Park*. Retrieved from https://crystalbrookenergypark.com.au/.

Neoen Australia. (2019c, November 4). *Hornsdale Power Reserve*. Retrieved from https://hornsdalepowerreserve.com.au/.

Pasqualetti, M., and Stremke, S. (2018). Energy Landscapes in a Crowded World: A First Typology of Origins and Expressions. *Energy Research and Social Science* 36 (2018): 94–105.

Planning South Australia. (2002). *Advisory Planning Notice*. Adelaide.

SA Planning Commission. (2019). *Discussion Paper on Propsoed Changes to Renewable Energy Policy in the Planning and Design Code*.

Scottish Natural Heritage, and the Countryside Agency. (2002). *Landscape Character Assessment Guidance for England and Scotland*. Scotland: Scottish Natural Heritage.

Sinclair, G. (2001). *The Potential Visual Impact of Wind Turbines in Relation to Distance: An Approach to the Environmental Assessment of Planning Proposals*. E.I. Services.

South Australian Government Department of Premier, and Cabinet. (2019, October 21). *Renewable Energy Atlas*. Retrieved from http://sagov.maps.arcgis.com.

State Planning Commission. (2019*). Discussion Paper on Proposed Changes to Renewable Energy Policy in the Planning and Design Code*.

Steffen, W., Grinevald, J., Crutzen, P., & McNeill, J. (2011). The Anthropocene: Conceptual and Historical Perspectives. *Philosophical Transactions of the Royal Society A: Mathematical, Physical and Engineering Sciences* 369 (1938): 842–867.

Stremke, S., and Dobbelsteen, A. V. D. (2013). *Sustainable Energy Landscapes: Designing, Planning, and Development*. Boca Raton, FL: CRC Press.

Thayer, R., and Hansen, H. (1989). *Consumer Attitude and Choice in Local Energy Development*. Davis, CA: Department of Environmental Design, University of California.

Tudor, C. (2014). *An Approach to Landscape Charater Assessment*. Peterborough, UK: Natural England.

United States Department of Agriculture. (1995). *Landscape Aesthetics A Handbook for Scenery Management, Agricultural Handbook Number 701*. Washington, DC: Untied States Department of Agriculture Forest Service.

University of Newcastle. (2002). *Visual Assessment of Wind Farms: Best Practice*. Scotland: Scottish Natural Heritage.

van den Dobbelsteen, Andy. (2014). *Urban Metabolism*, cited in Zeunert, 2017: 157.

Wax Design & Brett Grimm Landscape Architect. (2018). *Twin Creek Wind Farm Landscape Character and Probable Visual Effect Assessment.* North Sydney, NSW: RES Australia Pty Ltd.

Weller, R. (2014). Stewardship Now?: Reflections on Landscape Architecture's Raison d'être in the 21st Century. *Landscape Journal* 33 (2): 85–108.

Western Australian Planning Commision Department of Planning and Infrastructure. (2007). *Visual Landscape Planning in Western Australia: A Manual for Evaluation, Assessment, Siting and Design.* Perth, WA: Western Australian Planning Commission.

Wilson, E. (2016). *Half-Earth: Our Planet's Fight for Life.* New York: Liveright.

Wolsink. M. (2000). Wind Power and the NIMBY-Myth; Institutional Capacity and the Limited Significance of Public Support. *Renewable Energy* 21 (1): 49–64.

Wood, Denis. (1988). Unnatural Illusions: Some Words About Visual Resource Management. *Landscape Journal* 7 (2): 192–205.

Zeunert, J. (2017). *Landscape Architecture and Environmental Sustainability: Creating Positive Change through Design.* Bloomsbury: London.

6

Turkey and Sustainable Development Goals: A Nexus Approach to Clean Energy and Climate Action

Çiğdem Pekar

Introduction

In September 2015 The United Nations (UN) 2030 Agenda for Sustainable Development was adopted and came into force on January 1, 2016. The 2030 Agenda which aims to "end poverty, hunger and inequality, take action on climate change and the environment, improve access to health and education, build strong institutions and partnerships, and more"[1] includes a set of 17 Sustainable Development Goals (SDGs) and 169 targets. The SDGs which were designed to be universal are adopted by 193 countries. The SDGs are built on the Millennium Development Goals (MDGs) which were adopted in 2000 aiming to reduce extreme poverty via achieving eight anti-poverty targets by 2015. With a more holistic approach, SDGs include a more ambitious agenda and issues ranging from climate change to health, education, gender equality, and peace and justice for all.

Ç. Pekar (✉)
Çanakkale Onsekiz Mart University, Çanakkale, Turkey
e-mail: cbilezikci@comu.edu.tr

E. V. Shabliy et al. (eds.), *Discourses on Sustainability*,
https://doi.org/10.1007/978-3-030-53121-8_6

Although each of the SDGs and targets represent different aspects of the concept of sustainable development and address specific issues, many of these goals and targets are directly or indirectly affecting each other. Thus, this interaction, nexus, or interlinkages between goals and targets can be seen as a "complex web of sustainable development."[2]

Turkey has always been a strong supporter of the UN 2030 Agenda for Sustainable Development and worked toward the integration of the SDGs into its national policies. Turkish Long-Term Strategy and Five-Year National Development Plans (NDPs) and the Voluntary National Reviews (VNR), submitted to the UN High-level Political Forum on Sustainable Development (HLPF), present the most comprehensive framework regarding the country's approach toward the sustainable development concept, the UN 2030 Agenda for Sustainable Development, and the related goals.

Turkey also adopts a holistic approach toward the SDGs as a whole. In its 2nd Voluntary National Report (VNR) presented at the annual UN High-Level-Political-Forum which was titled "Strong Ground towards Common Goals," Turkey addresses major interactions between the SDGs and provides an SDG map of interaction which not only looks at the relations between the SDGs and their targets but also identifies direction and level of each relation. The report states that, with regard to the target that is being reviewed, while "impacted" refers to the positive or negative impact of another target, "impacting" refers to the effect and decisiveness of the target being reviewed on another target. Some targets both impact and are impacted by other targets. This is described as "mutual interaction."[3]

SDG 7 has clear linkages with the other 16 SDGs, particularly with SDG 13, which focuses on combatting climate change by emission reduction; SDG 7 mainly focuses on ensuring access to clean energy, increasing the share of renewables, and improving energy efficiency, which would have impacts on climate change and its impacts.

Turkey's policies in the framework of SDG 7 and SDG 13 are mostly mutually affecting each other. A clear example of Turkey's nexus approach to clean energy and climate action in the framework of SDG 7 and SDG 13 interaction can be seen in the overlapping national policies and common national institutions, which are responsible for SDGs

7 and 13. Turkey's nexus approach to clean energy and climate action is also obvious in its 7th National Communication to the UNFCCC dated 2018. Turkey defines its policy as "becoming a country fully integrating climate change policies with its development policies, disseminating energy efficiency, increasing the use of clean and renewable energy resources, actively participating in the efforts for tackling climate change within its special circumstances and providing its citizens with a high quality of life and welfare with low-carbon intensity."[4]

According to the Sustainable Development Solutions Network's "Sustainable Development Report 2019,"[5] Turkey's SDG Global rank is 79 out of 162 countries with a global index score of 68.5. Turkey's SDG 7—Affordable and Clean Energy score is 89.2 while significant challenges remain[6] and SDG 13—Climate Action score is 89.9 while major challenges remain.[7]

Turkey's General Approach Toward Sustainable Development Goals

Turkey's national policy toward the sustainable development concept and the country's priorities would be traced by a detailed reading of the country's National Development Plans (NDPs) since the Rio conference. NDPs, which can be regarded as the guiding document for the national policies and country's priorities, not only serve as the agenda-setting tools but also as the monitoring tool for the implementation of SDGs. The 7th NDP, which covers the 1996 and 2000 period, addresses the major topics discussed at the UN Rio Conference on Environment and Development in 1992.[8] After this date, the sustainable development concept was presented in Turkey's every NDP, which prioritized policy areas and major goals for the country. For instance, while the 7th NDP covers environment and economy integration policies, the 8th NDP (2001–2005) underlines a sustainable development approach that integrates the achieving economic development by protecting human health and the environment.[9] Under the "objectives, principles and policies" subtitle, 8th NDP puts that, in line with sustainable development approach "the main aim is to ensure economic and social development by protecting human health, ecological balance and cultural, historical and aesthetical values."[10]

It is observed that the 8th NDP's integrated approach is improved in the 9th NDP (2007–2013). The 9th Plan underlines this perspective and mentions Turkey will realize its development goals "in economic, social, and cultural areas in an integrated approach."[11] The Plan also adopts "development axes" in order to reach these strategic goals. These axes are: (a) Increasing Competitiveness, (b) Increasing Employment, (c) Strengthening Human Development and Social Solidarity, (d) Ensuring Regional Development, and (e) Increasing Quality and Effectiveness in Public Services. Related policies and country's priorities are to be considered under these axes.[12] In its 9th NDP Turkey also mentions the principle of "common but differentiated responsibilities" for the first time in one of its development plans and states that "fulfillment of international obligations will be realized in the framework of the principle of sustainable development and the principle of common but differentiated responsibility."[13]

The 10th NDP (2014–2018) adopts an integrated and multidimensional view on the country's economic and social development by emphasizing the strategies toward human-oriented development, the sustainability of resources, international competition power, and environmental protection.[14] The importance of "green growth" concept is also underlined as an environment-friendly development approach in the areas such as energy, agriculture, construction, transportation, and urbanization.[15] The plan recognizes the climate change and its impacts over different sectors of economy and in the framework of Turkey's national circumstances reemphasizes the concepts of "common but differentiated responsibilities" and "respective capabilities" in combatting climate change.

The 11th NPD (2019–2023) is the first development plan of the new Turkish Presidency Government System which was adapted in 2018. In this Plan, five fundamental axes are put in order to realize the vision of "strong Turkey." These axes are as the following: (a) stable and strong economy, (b) competitive production and efficiency, (c) qualified human beings and strong society, (d) livable cities, sustainable environment, and rule of law, and (e) democratization and good governance.[16]

Turkey also takes part in the voluntary national review (VNR) process, which is directed by the UN high-level political forum (HLPF). The VNR process not only aims to review member states' implementation of the 2030 Agenda but also to support its members by enabling experience sharing among them to strengthen policies and government institutions.[17] Turkey has recently submitted its 2019 report, which reviews the country's progress made in the 2010–2018 period toward realizing the SDGs.[18]

Turkey's Clean Energy Policy in the SDG 7 Context: Ensure Access to Affordable, Reliable, Sustainable, and Modern Energy for All

SDG 7 focuses on access to affordable, reliable, sustainable, and modern energy for all people. Five targets were set under the framework of SDG 7. These targets are:

(7.1) By 2030, ensure universal access to affordable, reliable, and modern energy services.
(7.2) By 2030, increase substantially the share of renewable energy in the global energy mix.
(7.3) By 2030, double the global rate of improvement in energy efficiency.

- (7.a) By 2030, enhance international cooperation to facilitate access to clean energy research and technology, including renewable energy, energy efficiency and advanced and cleaner fossil fuel technology, and promote investment in energy infrastructure and clean energy technology;
- (7.b) By 2030, expand infrastructure and upgrade technology for supplying modern and sustainable energy services for all in developing countries, in particular least developed countries, small island developing States, and land-locked developing countries, in accordance with their respective programs of support.[19]

According to the UN 2019 SDG Report, nearly 9 out of 10 people have access to electricity, and more people are using electricity than ever before. Furthermore, "access to electricity in poorer countries has begun to accelerate, energy efficiency continues to improve, and renewable energy is making impressive gains in the electricity sector."[20] According to the same report, although the share of renewable energy in total final energy consumption reached 17.5% in 2016, up from 16.6% in 2010, "progress in the electricity sector must extend to transportation and heating to meet an ambitious renewable energy target."[21] Regarding the energy efficiency, although there is a significant improvement, "more concerted action is needed to reach the SDG target."[22] Although there is a significant progress in the recent years, today three billion people continue to lack access to clean cooking solutions and there would still be 674 million people living without access to electricity in 2030.[23]

In Turkey's case, 100% of Turkey's population has access to electricity.[24] However, in order to fully achieve SDG 7, more actions particularly more investment in renewable energy resources is needed. Turkish Ministry of Energy and Natural Resources is the main coordinator for realizing the country's SDG 7 targets. Other responsible or relevant institutions are the following: Ministry of Environment and Urbanization; Ministry of Foreign Affairs; Ministry of Treasury and Finance; Ministry of National Education; Ministry of Industry and Technology; Ministry of Agriculture and Forestry; Ministry of Commerce; Ministry of Transportation and Infrastructure; General Directorate of State Hydraulic Works (DSI); Energy Market Regulatory Authority; Scientific and Technological Research Council of Turkey (TUBITAK); Turkish Cooperation and Coordination Agency (TIKA).[25]

Turkey states the country's general policy framework regarding SDG 7 in its 2nd VNR in 2019. According to the report, Turkey aims to:

(a) Maximize the use of domestic and renewable energy resources, and regain these resources in a secure, economic and qualified manner to economy;
(b) Minimize wastage and environmental effect of energy;
(c) Increase the share of renewable energy in power generation through resource diversification;

(d) Make efforts toward energy generation through resources such as hydroelectricity, along with solar, wind, geothermal, biomass;
(e) Subsidize domestic equipment use in renewable energy generation;
(f) Reduce energy intensity of industry;
(g) Provide supports for energy efficiency in buildings;
(h) Ease the burden of energy costs on economy;
(i) Scale up high efficiency motors in industry;
(j) Increase electrical vehicles in transport.[26]

In addition to Turkish NDPs as the primary guidelines for realizing the SGDs, there are several strategic policy documents prepared by several public institutions in Turkey. Furthermore, via guidance of the related strategies, public institutions are working on several legislations in order to provide access to affordable, reliable, sustainable, and modern energy for the society. SDG 7 targets and Turkey's leading strategies and plans regarding this target are summarized in Table 6.1.

It is clear from the table that Turkey's clean energy policy covers policies, strategies, and plans for several different sectors.

In its 2nd VNR, Turkey emphasizes its national policy priorities and lists "four focus areas" regarding SDG 7, which the country will align its policies with, for achieving the 2030 Agenda.[27] These four focus areas are as the following: (a) access to energy for all, (b) increasing renewable energy share in total supply, (c) improving energy efficiency, and (d) development of clean energy technologies and infrastructure.

(a) **"Access to energy for all"**: Turkey aims to increase the share of the private sector in electricity production and commerce in order to provide energy. To this end, not only several state-owned energy power plants were privatized since the 2000s but also private sector mostly took over the natural gas distribution in the country. Furthermore, renewable energy power plants' integration to the national energy mix is accelerated.
(b) **"Increasing renewable energy share in total supply"**: Apart from providing access to energy, Turkey is working toward increasing

Table 6.1 SDG 7 targets and related Turkish strategies/plans/documents

Targets	Strategy/plan/document
(7.1) (7.2) (7.3)	Ministry of Energy and Natural Resources Strategic Plan (2015–2019)[a]
(7.1) (7.2)	Electrical Energy Market and Supply Security Strategy (2009)[b]
(7.1)	Electrical Energy Sector Reform and Privatization Strategy (2004)[c]
(7.1) (7.2)	Turkish Electricity Transmission Corporation Strategic Plan (2015–2019)[d]
(7.1)	Turkish Coal Corporation Strategic Plans (2015–2019 and 2019–2023)[e]
(7.2)	Turkey National Renewable Energy Action Plan (2014–2019)[f]
(7.2)	Turkey National Climate Change Strategy (2010–2023)[g]
(7.2)	Turkey's National Climate Change Adaptation Strategy and Action Plan (2011–2023)[h]
(7.3)	Energy Efficiency Strategy (2012–2023)[i]
(7.3)	Turkey National Energy Efficiency Action Plan (2017–2023)[j]

[a] Ministry of Energy and Natural Resources (MENR), Enerji ve Tabii Kaynaklar Bakanlığı Stratejik Planı (2015–2019), https://enerji.gov.tr/File/?path=ROOT%2f1%2fDocuments%2fStratejik%20Plan%2fETKB%202015-2019%20Stratejik%20Plani.pdf (Last accessed October 21, 2019)
[b] General Directorate of Energy Affairs, Electrical Energy Market and Supply Security Strategy, Elektrik Enerjisi Piyasası ve Arz Güvenliği Strateji Belgesi (2009), https://www.eigm.gov.tr/File/?path=ROOT%2f4%2fDocuments%2fEnerji%20Politikas%C4%B1%2fElectricity%20Market%20and%20Security%20of%20Supply%20Strategy%20Paper.pdf (Last accessed October 21, 2019)
[c] General Directorate of Energy Affairs, Electrical Energy Sector Reform and Privatization Strategy, Elektrik Enerjisi Sektörü Reformu ve Özelleştirme Stratejisi Belgesi (2004), https://www.eigm.gov.tr/File/?path=ROOT%2f4%2fDocuments%2fEnerji%20Politikas%C4%B1%2fElektrik%20Enerjisi%20Sekt%C3%B6r%C3%BC%20Reformu%20ve%20%C3%96zelle%C5%9Ftirme%20Strateji%20Belgesi.pdf (Last accessed October 21, 2019)
[d] Turkish Electricity Transmission Corporation, TEİAŞ Stratejik Planı (2015–2019), https://www.teias.gov.tr/tr-TR/stratejik-plan (Last accessed October 21, 2019)
[e] Turkish Coal Corporation, TKİ Stratejik Planı (2015–2019), http://www.tki.gov.tr/depo/2017/TKi_StratejikPlan.pdf, TKİ Stratejik Planı (2019–2023), http://www.tki.gov.tr/depo/file/TK%C4%B0_2019_2023_Stratejik_Plan.pdf (Last accessed October 21, 2019)
[f] General Directorate of Energy Affairs, National Renewable Energy Action Plan for Turkey (NREAP), Yenilenebilir Enerji Eylem Planı (2014–2019), https://www.eigm.gov.tr/File/?path=ROOT%2f4%2fDocuments%2fEnerji%20Politikas%C4%B1%2fNational_Renewable_Energy_Action_For_Turkey.pdf (Last accessed October 21, 2019)
[g] Turkey National Climate Change Strategy, Türkiye İklim Değişikliği Stratejisi (2010–2023), https://www.gmka.gov.tr/dokumanlar/yayinlar/Turkiye-Iklim-Degisikligi-Stratejisi.pdf (Last accessed October 21, 2019)
[h] Turkey's National Climate Change Adaptation Strategy and Action Plan, Türkiye'nin İklim Değişikliği Uyum Stratejisi ve Eylem Planı (2011–2023), https://webdosya.csb.gov.tr/db/iklim/editordosya/uyum_stratejisi_eylem_plani_TR.pdf (Last accessed October 21, 2019)
[i] Ministry of Energy and Natural Resources (MENR), Enerji Verimliliği Strateji Belgesi (2012–2023), https://www.resmigazete.gov.tr/eskiler/2012/02/20120225-7.htm (Last accessed October 25, 2019)
[j] Turkey National Energy Efficiency Action Plan, Ulusal Enerji Verimliliği Eylem Planı (2017–2023), http://www.yegm.gov.tr/document/20180102M1_2018.pdf (Last accessed October 21, 2019)

the country's renewable energy production for several reasons. Via having more renewable energy generation, Turkey not only aims to diversify its energy resources and to reduce its dependency on energy imports but also aims to reduce its greenhouse gas emissions (GHGs). As it is stated in most of the national strategy plans, by the year of 2023, Turkey aims to achieve at least 30% share of renewable energy in its total electricity generation. To this end, Renewable Energy Resources Support Mechanism (YEKDEM) was established in 2011 to regulate the renewable energy tariff supports (Feed-In Tariffs) and guaranteed purchase tariff prices in order to increase the investment in renewable energy sectors.[28] Furthermore, domestically manufactured equipment and machinery support mechanism is established which provided price subsidy if domestically manufactured machinery and equipment is used in renewable energy facility. All of these incentive mechanisms are regarded as significant tools to support domestic and foreign renewable energy investors.

Furthermore, the 2nd Turkey VNR provides detailed data on the country's progress in the generation of electricity via renewable energy resources and makes a comparison with Organization for Economic Cooperation and Development (OECD) countries. The report states that the installed power of Turkey's renewable energy rose to 42.5 GW and the share of renewable energy increased to 32.5% by the end of 2018 (Figs. 6.1 and 6.2).[29]

(c) **"Improving energy efficiency"**: 2017 Turkish National Energy Efficiency Action Plan presents the strategy and targets of Turkey in the framework of improving energy efficiency, describes actions for all sectors and defines responsibilities for institutions. According to this plan a cumulative 23.9 Mtoe of primary energy consumption reduction is targeted in the period of 2017–2023.[30] That number equals to a 14% decrease in primary energy consumption for this period and in order to achieve this target 10.9 billion dollar investment is foreseen.[31] Furthermore, the plan states that as a significant indicator of energy efficiency, primary energy intensity index decreased by 23.1% between 2000 and 2015.[32]

Source	Generation 2013	Generation 2017	Ranking 2017	Ranking 2017	Comparison Group
Hydroelectricity	59.4 TWh	58.2 TWh	7	6	OECD
Geothermal	1.4 TWh	6.1 TWh	7	4	OECD
Solar	0.017 TWh	2.9 TWh	27	14	OECD
Wind	7.6 TWh	17.9 TWh	12	7	OECD

Fig. 6.1 Turkey's generation of electricity via renewable energy resources (*Source* Turkey's 2nd VNR 2019 Sustainable Development Goals, "Strong Ground towards Common Goals," https://sustainabledevelopment.un.org/content/documents/23862Turkey_VNR_110719.pdf, 78)

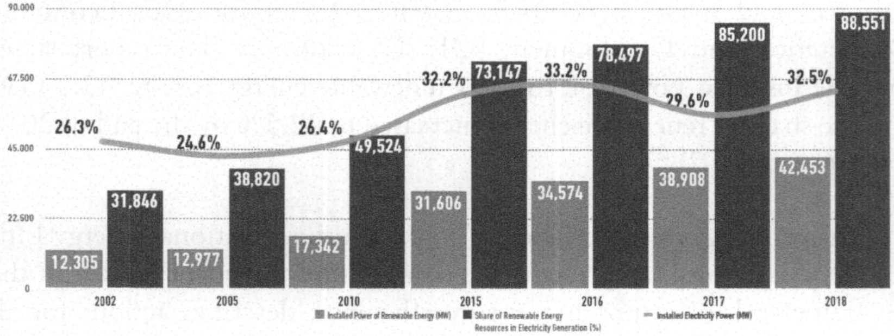

Fig. 6.2 Share of renewable energy resources in electricity generation (%) (*Source* Turkey's 2nd VNR 2019 Sustainable Development Goals, "Strong Ground towards Common Goals," https://sustainabledevelopment.un.org/content/documents/23862Turkey_VNR_110719.pdf, 80)

As Turkey states in its 2nd VNR, "in the context of the National Energy Efficiency Action Plan, approximately 1 million TOE of energy saved through 1.2 billion USD investment up to date. The loss in the grid recently reduced to 12% through energy efficiency efforts and expanding distributed generation."[33]

(d) **"Development of clean energy technologies and infrastructure"**: While Turkey is actively taking part in international renewable energy collaborations, the country is also working ambitiously toward establishment of new national R&D facilities which would enable the country to develop its own clean energy infrastructure.

Renewable Energy Resource Areas (YEKA) is a key incentive mechanism and implementation tool in this regard. According to the YEKA regulation which was published in 2016 in the Official Gazzette, the aim of establishing YEKAs is to benefit from renewable energy resources "effectively and efficiently."[34] Furthermore, a significant contribution to the domestic renewable energy technology production and support for the country's R&D capacity and technology transfer is foreseen by the establishment of YEKA mechanism.

Turkey's Climate Action Policy in the SDG 13 Context: Take Urgent Action to Combat Climate Change and Its Impacts

According to UN 2019 SDG Report, "the global mean temperature for 2018 was approximately 1 °C above the pre-industrial baseline, and the last four years have been the warmest on record."[35] This fact is also underlined in several UN documents. According to UN Secretary General's 2019 report titled "Progress towards the Sustainable Development Goals":

> in 2017, greenhouse gas concentrations reached new highs, with globally averaged mole fractions of CO_2 at 405.5 parts per million (ppm), up from 400.1 ppm in 2015, and at 146% of pre-industrial levels.[36]

In his report, UN Secretary-General also states that "with rising greenhouse gas emissions, climate change is occurring at rates much faster than anticipated and its effects are clearly felt worldwide."[37]

In Turkey's case, as it is stated in the country's 7th National Communication to the UNFCCC, a large part of the country is located in the dry summer subtropical Mediterranean climate zone which puts Turkey among the "medium-high risk countries in terms of both present climate, climate change and variability, and future climate."[38] Several extreme weather situations which have led to floods and droughts in the country are clear examples of how Turkey is affected by climate change. Furthermore, the country is also experiencing climate change impact on land and water resources, which have direct effects on the public's life.

In order to answer this global problem, SDG 13 focuses on taking urgent action to combat climate change and its impacts. "Acknowledging that the United Nations Framework Convention on Climate Change (UNFCCC) is the primary international, intergovernmental forum for negotiating the global response to climate change," the UN has defined five targets for SDG 13. These are:

(13.1) Strengthen resilience and adaptive capacity to climate-related hazards and natural disasters in all countries.
(13.2) Integrate climate change measures into national policies, strategies, and planning.
(13.3) Improve education, awareness-raising and human and institutional capacity on climate change mitigation, adaptation, impact reduction and early warning.

   (13.a) Implement the commitment undertaken by developed-country parties to the United Nations Framework Convention on Climate Change to a goal of mobilizing jointly $100 billion annually by 2020 from all sources to address the needs of developing countries in the context of meaningful mitigation actions and transparency on implementation and fully operationalize the Green Climate Fund through its capitalization as soon as possible.

(13.b) Promote mechanisms for raising capacity for effective climate change-related planning and management in least developed countries and small island developing States, including focusing on women, youth and local and marginalized communities.[39]

It is clear that SDG 13 does not directly cover indicators such as greenhouse emissions and carbon intensity of the economy. As it is emphasized, that is because climate action is monitored under the framework of the Paris Agreement rather than the sustainable development agenda. Although this situation would be regarded as a factor to decrease the effectiveness of SDGs in terms of climate action, it is also clear that targets and indicators of SDG 13 foresee the implementation of the Paris Agreement.

According to the World Bank classification, Turkey is a member of the upper-middle-income country group.[40] In 2004, Turkey became a party to the UNFCCC as an Annex I Party with "special circumstances," which puts Turkey in a situation different from that of other Parties, which are part of Annex I to the Convention. Thus, since that date, Turkey has been emphasizing the principles of "common but differentiated responsibilities" and "respective capabilities" in its national and international policy documents. Turkey does not have the obligation to work toward 13.a and 13.b targets.

Following the 21st Conference of the Parties (COP21) in 2015, Turkey also submitted its first intended nationally determined contribution (INDC) in October 2015 to the UNFCCC. In its INDC Turkey undertakes "up to 21% reduction in GHG emissions from the Business as Usual (BAU) level by 2030."[41] As it is stated in its 2nd VNR, Turkey is "responsible for 0.7% of total global emissions since the industrial revolution" and working toward to achieve its INDC target via its climate action policy, which is formed by its public institutions as a result of intensive consultation with all the stakeholders.

At the national basis, Turkish Ministry of Environment and Urbanization is the main coordinator for SDG 13. Other responsible or relevant Institutions are: Ministry of Energy and Natural Resources, Ministry of

Treasury and Finance, Ministry of National Education, Ministry of Agriculture and Forestry, Ministry of Commerce, Turkish Cooperation and Coordination Agency (TIKA), Disaster and Emergency Management Authority (AFAD), Turkish Natural Catastrophe Insurance Pool.[42]

Turkey states the country's general policy framework regarding SDG 13 in its 2nd VNR in 2019. According to the report Turkey's major goals are to

(a) Contribute to combat global climate change by emission reduction and adaptation in line with the principles of common but differentiated responsibilities and respective capabilities,
(b) Increase the share of domestic and renewable energy sources by considering supply security,
(c) Increase energy efficiency, reducing losses and illegal use of electricity,
(d) Prioritize transport systems that provide energy efficiency and use of clean fuel and environmentally friendly vehicles,
(e) Encourage railway and maritime transport and developing combined transport opportunities,
(f) Develop building standards for safe, energy efficient, environmentally friendly and expand energy efficiency practices in buildings,
(g) Strengthen the integration of transportation infrastructure with other infrastructures by encouraging public transport in urban transportation,
(h) Making use of green growth opportunities in areas such as energy, industry, agriculture, transportation, construction, services and urbanization; promote new business areas, R&D and innovation that provide environmentally conscious economic growth,
(i) Promoting recycling and reuse through the implementation of integrated waste management principles,
(j) Taking necessary measures to provide water saving in basins, to combat against drought and to prevent pollution by evaluating the effects of climate change and all activities in water basins on water quantity and quality,
(k) Ensuring conservation, sustainable use and efficient management of water resources,

(l) Combatting against desertification and erosion effectively, monitoring the environmental and social impacts of agricultural activities on soil resources and increasing preventive measures,
(m) Strengthening management capacity and mechanisms for disaster risk reduction.[43]

In addition to Turkish NDPs for realizing SDG 13, there are several strategic policy documents prepared by different public institutions which are working on several legislations in order to take urgent action to combat climate change and its impacts. SDG 13 targets and Turkey's leading strategies and plans regarding these targets can be summarized as the following (Table 6.2).

Turkey's climate change policy is constructed in the framework of several cross-cutting and sectoral policies, strategies, and plans. The main policy documents that are dedicated only to Turkey's policies in the context of SDG 13 are National Climate Change Strategy (NCCS) and National Climate Change Action Plan. These documents also have a significant role in the country's SDG 7 policies.

In its 2nd VNR, Turkey emphasizes its national policy priorities in regard to SDG 13. In this framework, "four focus areas" are listed, which the country will align its policies with, for achieving the 2030 Agenda. These focus areas are listed as the following: (a) adaptation to climate change, (b) mainstreaming climate action into policies, (c) awareness-raising and capacity increase, and (d) climate finance.

(a) **Adaptation to climate change**: Climate change has been strongly affecting Turkey in the recent years. Particularly floods are causing significant economic losses. According to the country's 2nd VNR, "it is foreseen that the main impacts of climate change on Turkey will be drought, desertification, decline in agricultural productivity, loss of biological diversity and ecosystem services, increase in forest fires, epidemics and pests."[44] Thus, it is clear that climate change adaptation and implementation of public institutions presents unique importance to the detection of risks, build resilience and take actions against these risks.

Table 6.2 SDG 13 targets and related Turkish strategies/plans/documents

Targets	Strategy/plan/document
(13.1)	National Earthquake Strategy and Action Plan (2012–2023)[a]
(13.1)	Turkey Disaster Response Plan (2013)[b]
(13.1)	National Drought Management Strategy and Action Plan (2017–2023)[c]
(13.1) (13.2) (13.3)	Climate Change Adaptation Strategy and Action Plan (2011–2023)[d]
(13.2)	Energy Efficiency Strategy Paper (2012–2023)[e]
(13.2) (13.3)	National Strategy and Action Plan to Combat Desertification (2015–2023)[f]
(13.2) (13.3)	Action Plan to Combat Erosion (2013–2017)[g]
(13.2)	National Biological Diversity Strategy and Action Plan (2007–2017)[h]
(13.2) (13.3)	National Strategy and Action Plan for Combatting Agricultural Drought (2018–2022)[i]
(13.2)	National Program and Action Plan to Decrease the Negative Effects of Climate Change on Health[j]
(13.2)	National Waste Management and Action Plan (2016–2023)[k]
(13.2) (13.3)	Turkey Disaster Response Plan (2013)[l]
(13.1) (13.2) (13.3)	National Climate Change Strategy (2010–2023)[m]
(13.1) (13.2) (13.3)	Turkey's National Climate Change Adaptation Strategy and Action Plan (2011–2023)[n]
(13.1) (13.2) (13.3)	Disaster and Emergency Management Authority (AFAD) Strategic Plan[o]

[a] Turkish National Earthquake Strategy and Action Plan, Ulusal Deprem Stratejisi ve Eylem Planı (2012–2023), https://www.resmigazete.gov.tr/eskiler/2011/08/20110818-13-1.pdf (Last accessed October 25, 2019)

[b] Turkey Disaster Response Plan, Türkiye Afet Müdahale Planı (TAMP) (2013), https://www.afad.gov.tr/kurumlar/afad.gov.tr/2419/files/Afet_Mud_Pl_ResmiG_20122013.pdf (Last accessed October 25, 2019)

[c] National Drought Management Strategy and Action Plan, Ulusal Kuraklık Yönetimi Strateji Belgesi ve Eylem Planı (2017–2023), https://www.tarimorman.gov.tr/SYGM/Belgeler/Ulusal%20Kurakl%C4%B1k%20Y%C3%B6netimi%20Strateji%20Belgesi%20ve%20Eylem%20Plan%C4%B1/Ulusal%20Kurakl%C4%B1k%20Y%C3%B6netimi%20Strateji%20Belgesi%20ve%20Eylem%20Plan%C4%B1.pdf (Last accessed October 25, 2019)

[d] Climate Change Adaptation Strategy and Action Plan, İklim Değişikliği Uyum Stratejisi ve Eylem Planı (2011–2023), https://webdosya.csb.gov.tr/db/iklim/editordosya/uyum_stratejisi_eylem_plani_TR.pdf (Last accessed October 21, 2019)

[e] Energy Efficiency Strategy Paper, Enerji Verimliliği Strateji Belgesi (2012–2023), https://www.enerji.gov.tr/File/?path=ROOT%2f1%2fDocuments%2fMevzuat.%2fEnerji_Verimliligi_Strateji_Belgesi_2012_2023.pdf (Last accessed October 25, 2019)

[f]National Strategy and Action Plan to Combat Desertification, Çölleşme ile Mücadele Ulusal Stratejisi ve Eylem Planı (2015–2023), https://www.tarimorman.gov.tr/CEM/Belgeler/collesme%20belgeleri%20arsiv/Sayfa04/%C3%87%C3%96LLE%C5%9EME_16X23.pdf (Last accessed October 25, 2019)
[g]Action Plan to Combat Erosion, Erozyonla Mücadele Eylem Planı (2013–2017), https://www.ogm.gov.tr/ekutuphane/Yayinlar/Erozyonla%20M%C3%BCcadele%20Eylem%20Plan%C4%B1.pdf (Last accessed October 25, 2019)
[h]National Biological Diversity Strategy and Action Plan, Ulusal Biyolojik Çeşitlilik Stratejisi ve Eylem Planı (2007–2017, http://www.milliparklar.gov.tr/dosyalar/nbap.pdf (Last accessed October 25, 2019)
[i]National Strategy and Action Plan for Combatting Agricultural Drought, Tarımsal Kuraklıkla Mücadele Stratejisi ve Eylem Planı (2018–2022), https://istanbul.tarimorman.gov.tr/Belgeler/KutuMenu/KuraklikEylemPlani/Kurakl%C4%B1k%20Kitab%C4%B1%20%C4%B0c%20Sayfalar.pdf (Last accessed October 25, 2019)
[j]National Program and Action Plan to Decrease the Negative Effects of Climate Change on Health, İklim Değişikliğinin Sağlık Üzerine Olumsuz Etkilerinin Azaltılması Ulusal Programı ve Eylem Planı, https://hsgm.saglik.gov.tr/depo/birimler/cevre-sagligi/2-ced/iklim-degisikligi/Iklim_Degisikligi_Eylem_Plani.pdf (Last accessed October 25, 2019)
[k]National Waste Management and Action Plan, Ulusal Atık Yönetimi ve Eylem Planı (2016–2023), https://webdosya.csb.gov.tr/db/cygm/haberler/ulusal_at-k_yonet-m--eylem_plan--20180328154824.pdf (Last accessed October 25, 2019)
[l]Turkey Disaster Response Plan, Türkiye Afet Müdahale Planı (TAMP) (2013), https://www.afad.gov.tr/kurumlar/afad.gov.tr/2419/files/Afet_Mud_Pl_ResmiG_20122013.pdf (Last accessed October 25, 2019)
[m]National Climate Change Strategy, Ulusal İklim Değişikliği Stratejisi (2010–2023), https://www.gmka.gov.tr/dokumanlar/yayinlar/Turkiye-Iklim-Degisikligi-Stratejisi.pdf (Last accessed October 21, 2019)
[n]Turkey's National Climate Change Adaptation Strategy and Action Plan, Türkiye'nin İklim Değişikliği Uyum Stratejisi ve Eylem Planı (2011–2023), https://webdosya.csb.gov.tr/db/iklim/editordosya/uyum_stratejisi_eylem_plani_TR.pdf (Last accessed October 25, 2019)
[o]Disaster and Emergency Management Authority (AFAD) Strategic Plan, AFAD Stratejik Planı (2013–2017), https://www.afad.gov.tr/kurumlar/afad.gov.tr/2402/files/afadsp2013-2017.pdf (Last accessed October 25, 2019)

(b) **Mainstreaming climate action into policies**: Turkey's international commitments to combat global climate change have guided the country's main climate action strategies and development plans. Turkey's NDPs have been addressing climate change action in several sectors, including energy, industry, transportation, agriculture, urbanization, and water resources. As Turkey outlined in its 2nd VNR, the country aims to combat climate change mainly by emission reduction, increase in renewable energy sources and use of green

> growth opportunities in several sectors in order to provide "environmentally conscious economic growth." Furthermore, Turkey has also made particular reference to the concept of "green growth" in the 10th NDP as an environment-friendly development approach.

In this context, in the foreword of its 7th National Communication to the UNFCCC dated 2018, Turkey defines its vision within the scope of National Climate Change Strategy as "becoming a country fully integrating climate change policies with its development policies, disseminating energy efficiency, increasing the use of clean and renewable energy resources, actively participating in the efforts for tackling climate change within its special circumstances and providing its citizens with a high quality of life and welfare with low-carbon intensity."[45]

There exist other particular documents, which aim to mainstream climate action into Turkey's policies. In this regard, Turkey's National Climate Change Strategy for the 2010–2023 period was adapted in 2010 and the National Climate Change Action Plan of Turkey for the 2011–2023 period was announced in 2012. These two policy documents present particular importance for Turkey's strategies and actions in mitigation, compliance, financing, technology policies related to energy. Turkey's "common but differentiated responsibilities" approach to control greenhouse gas emissions is emphasized in both documents.

Energy Efficiency Strategy Document[46] which was announced in 2012 for the 2012–2023 period is also another important national policy document regarding Turkey's energy efficiency policy. According to this strategy, Turkey aims to reduce the country's energy intensity by 20% compared to its level in 2011 by 2023. Furthermore, Turkey's National Energy Efficiency Action Plan (NEEAP), which covers 2017–2023 period updates this target as 14% reduction in primary energy consumption.[47]

Regarding the renewable energy generation and targets, Turkey's National Renewable Energy Action Plan (NREAP) adopted in 2014 can be regarded as a comprehensive document. The Plan targets an increase of 30% in the share of renewable energy by 2023 (61 GW in 2023).[48] In its 2nd VNR, Turkey states that, "even in 2017, which was dry and set record highs in gas consumption, the share of renewable energy resources increased to 29.7%."[49] Turkey also emphasizes the importance of incentive mechanisms to increase the use of renewable energy resources such as the establishment of YEKDEM and its feed-in tariffs system.

c) **Awareness-raising and capacity increase**: Turkey is not only working toward improving national institutional capacity on climate change mitigation but also to improve public awareness and participation in climate action at various levels of education and awareness-raising activities. Turkey aims to increase the public's sensitivity to climate change and include them in its climate actions through the organization of training, education programs, and publications by local administrations and civil society. Disaster and Emergency Management Presidency (AFAD) is the leading public institution to arrange such activities.
d) **Climate finance**: UNFCCC system distinguishes its parties as Annex I, Annex II, and Non-Annex I countries according to their commitments. While the Annex I countries are industrialized countries (members of the OECD and Economies in Transition Countries [EIT]), Annex II countries are OECD members of Annex I, but not the EIT Parties who "are required to provide financial resources to enable developing countries to undertake emissions reduction activities under the Convention and to help them adapt to adverse effects of climate change."[50]

Non-Annex I countries consist of developing countries. Although Turkey is one of the founding members of the OECD, as a developing country, it submitted its demand to be removed from Annexes I and II. As a result of its application, the country was removed from Annex II but not from Annex I. Furthermore, parties of the Convention are invited to recognize the "special circumstances of Turkey, which place Turkey, after becoming a Party, in a situation different from that of other Parties included in Annex I to the Convention."[51] In 2004, Turkey became a Party to the UNFCCC as an Annex I Party with "special circumstances." Thus, Turkey does not have the obligation to provide financial support to developing countries. In its 2nd VNR, Turkey underlines the country is receiving financial resources such as EU funds and multilateral climate funds from multilateral development banks, bilateral development agencies, funds, such as Clean Technology Fund and Global Environment Facility (GEF) since it is a developing country itself.[52] Furthermore, the report states that Turkey has generated "over 4 million USD of income by trading 3.2 million tons of carbon credit in voluntary markets in 2015."[53]

Nexus Between Clean Energy and Climate Action in the Context of Turkey

Achieving clean energy in the framework of SDG 7 has clear linkages with SDG 13, since it is mainly focusing on increasing the share of renewables and improving energy efficiency which would contribute to the SDG 13 targets for combatting global climate change by emission reduction. Turkey's GHG emission projections and performance are at the center of clean energy-climate action nexus.

Today most of the world's energy is produced from fossil fuels, which are the major contributor to global GHG emissions. That is also a fact for Turkey. In 2016, the share of fossil fuels in Turkey's total primary energy supply was 87.3%. Coal (33.7%) and natural gas (32.5%) had the highest share in total electricity generation in 2016. It is also stated in the 7th National Communication that energy sector is the major source of Turkish GHG emissions. As the report states it,

> fuel combustion emissions are the major source of energy sector GHG emissions, and responsible for 97.7% of the whole energy sector GHG emissions. Energy industries were the main contributor of GHG emissions from fuel combustion with 41%. It is followed by transportation sector with 23.2%, manufacturing industries with 16.9%, residential, commercial and institutional sector with 16.1% and, agriculture, forestry and fishing sector with 2.8%.[54]

Furthermore, in its 7th National Communication Turkey underlines not only the increase in its population but also in the country's GDP and energy demand for 1990–2016. According to the report, while final energy consumption of the country has increased from 42.2 Mtoe in 1990 to 104.5 Mtoe in 2016, the total increase of GHG emissions over the same period by 135.4%.[55]

It is clear that on the national level, although Turkey is not a party to the Paris Climate Agreement, achieving SDG 7 and its targets would serve the country's efforts in climate action. Particularly target 7.2, which foresees the increase in renewable energy means a decrease in the usage of fossil fuels in its energy mix and de-carbonization of energy generation. Less fossil use would significantly contribute to the SDG 13, which focuses on climate action and emission reduction. Replacing fossil by renewable energy would contribute to major reductions in Turkey's GHG emissions. Furthermore, achieving SDG 7 and expanding the share of renewables in the energy mix would enable Turkey to implement its green growth approach.

SDG 13 and its targets would enable Turkey to integrate its renewable energy and energy efficiency measures into national policies and planning more coherently. Furthermore, improving public awareness through education and training activities would enable to improve the human and national institutional capacity on climate change mitigation. In this regard, international funds are being used through projects, national universities are initiating renewable energy education in bachelors and graduate degrees and an increased number of academic theses are being written on this topic.

On the other hand, although Turkey has ambitious plans to increase the renewable energy resources in its energy mix, an increase in coal production for energy security is on the national agenda. In MENR Strategic Plan (2015–2019) it is stated that in order to utilize from national energy resources, Turkey plans to increasingly use the national coal resources for electricity generation in order to meet increasing energy demand by increasing the annual electricity generation from domestic coal.

Furthermore, in the framework of National Energy and Mining Policy, which was announced in 2017, Turkey declared its intension to increasingly utilize its national coal resources for energy generation in order to decrease the country's dependency on imports. By 2018, in Turkey 113.3 TWh electricity is generated from coal-based plants, which is equal to 37.3% of total electricity generation.[56] As of 2019, Turkey has 67 coal-fired power units, which has a total of 16.3 GW capacity and more are currently under development.[57] It is clear that Turkey's plans for the increased use of coal for electricity presents an irony with the country's emission targets.

An in-depth look at the country's GHG emission data and UNFCCC report on Turkey's 6th National Communication indicate this point. UNFCCC report states that Turkey's declaration of its aim "to increase local coal based electricity power production from 32 billion kWh in 2013 to 57 billion kWh in 2018, even though this is likely to lead to an increase in emissions."[58]

The 7th National Communication Turkey's projections of GHGs by 2030 are based on two scenarios: BAU Scenario (Without Measures) and Mitigation (With Measures) Scenario. In its 7th National Communication Turkey puts that while the Mitigation Scenario suggests approximately 246 Mton CO_2 equivalent of emission reduction compared to BAU Scenario by the year 2030 which corresponds to up to 21% reduction.[59] According to the latest GHG inventory of Turkey, the report states that:

> total GHG emissions were 496.1 Mt of CO_2 equivalents (CO_2 eq.) excluding the LULUCF sector and 428.0 Mt CO_2 eq. including the LULUCF sector in 2016. This represents 135.4 increase as compared to 1990 level. In overall 2016 GHG emissions without LULUCF, the energy sector had the largest portion with 72.8%.[60]

As it has been stated, Turkey's GHG emission projections and performance are at the center of clean energy–climate action nexus. Turkey's GHG emission projections in both scenarios can be seen in the Fig. 6.3.

Conclusion

Turkey has always been a strong supporter of the UN 2030 Agenda for Sustainable Development and worked toward integration of SDGs into its national policies. Turkey has adopted and successfully integrated the

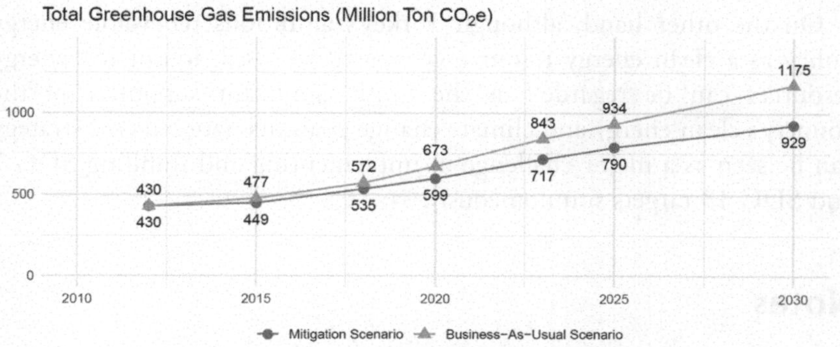

Fig. 6.3 Turkey's GHG emission projections in business as usual scenario and with measures scenario (*Data Source* UNFCCC, 2018, "Turkey's 7th National Communication," https://unfccc.int/sites/default/files/resource/496715_Turkey-NC7-1-7th%20National%20Communication%20of%20Turkey.pdf, 121 [Last accessed September 18, 2019])

concept of "sustainable development" into its NDPs, laws, strategies, and policy documents since the 1992 UN Conference on Environment and Development.

An increase in the use of renewable energy (target 7.2) and increased energy efficiency (target 7.3) have an impact on the environment. This increase would not only serve for the main goal of ensuring access to clean energy for all but also is a critical tool in strengthening resilience and adaptive capacity to climate-related hazards (target 13.1) and combat with global climate change and GHG emissions. Turkey is working toward achieving these interacting targets simultaneously.

Turkey also adopts a holistic approach toward the SDGs and associated SDG targets as interacting with each other. In this context, a strong energy–climate nexus of interactions is especially obvious among SDG 7 and SDG 13. Especially, clean energy and climate change topics will be significantly important for Turkey in the upcoming years. Turkey has ambitious policies for ensuring access to clean energy by focusing on increasing the share of renewables and improving energy efficiency in the framework of SDG 7 and sees this policy as a major contribution to achieve SDG 13 targets for combatting global climate change by emission reduction.

On the other hand, although Turkey's ambitious renewable energy policy as a clean energy resource compared to fossil dominated energy resources can be regarded as the most significant supporter of the country's clean energy and climate change goals, its national coal strategy can be seen as a major challenge in implementing and fulfilling SDG 7 and SDG 13 targets simultaneously.

Notes

1. United Nations Development Programme (UNDP), "World Leaders Adopt Sustainable Development Goals," September 25, 2015, https://www.undp.org/content/undp/en/home/presscenter/pressreleases/2015/09/24/undp-welcomes-adoption-of-sustainable-development-goals-by-world-leaders.html (Last accessed August 2, 2019).
2. For further information on the IAEG-SDGs working group, which is working on interlinkages between SDG indicators and establishing the global SDG indicator framework, see; United Nations Statistics Division, Sustainable Development Goal indicators website, https://unstats.un.org/sdgs/files/Working-Group-ToR--Interlinkages.pdf (Last accessed September 18, 2019).
3. Turkey's 2nd VNR 2019 Sustainable Development Goals, "Strong Ground towards Common Goals," https://sustainabledevelopment.un.org/content/documents/23862Turkey_VNR_110719.pdf, 29 (Last accessed October 15, 2019).
4. UNFCCC, 2018, "Turkey's 7th National Communication," https://unfccc.int/sites/default/files/resource/496715_Turkey-NC7-1-7th%20National%20Communication%20of%20Turkey.pdf (Last accessed September 18, 2019).
5. Bertelsmann Stiftung and Sustainable Development Solutions Network (SDSN), Sustainable Development Report 2019, https://www.sdgindex.org/reports/sustainable-development-report-2019/ (Last accessed October 29, 2019).
6. Which means "score stagnating or increasing at less than 50% of required rate." Ibid.
7. Which means "score decreasing." Ibid.
8. Presidency of Turkey, Presidency of Strategy and Budget, "Turkey's 7th National Development Plan," http://www.sbb.gov.tr/wp-content/

uploads/2018/11/Yedinci-Be%C5%9F-Y%C4%B1ll%C4%B1k-Kalk%C4%B1nma-Plan%C4%B1-1996-2000%E2%80%8B.pdf (Last accessed October 15, 2019).

9. Presidency of Turkey, Presidency of Strategy and Budget, "Turkey's 8th National Development Plan," http://www.sbb.gov.tr/wp-content/uploads/2018/11/Eight-Five-Year-Development-Plan-2001-2005.pdf (Last accessed October 15, 2019).
10. Ibid., 212.
11. Presidency of Turkey, Presidency of Strategy and Budget, "Turkey's 9th National Development Plan," http://www.sbb.gov.tr/wp-content/uploads/2018/11/Ninth_Development_Plan_2007-2013.pdf, 11 (Last accessed October 15, 2019).
12. Ibid.,12.
13. Ibid., 8.
14. Presidency of Turkey, Presidency of Strategy and Budget, "Turkey's 10th National Development Plan," http://www.sbb.gov.tr/wp-content/uploads/2018/11/Onuncu-Kalk%C4%B1nma-Plan%C4%B1-2014-2018.pdf, 1 (Last accessed October 15, 2019).
15. Ibid., 137.
16. Presidency of Turkey, Presidency of Strategy and Budget, "Turkey's 11th National Development Plan," http://www.sbb.gov.tr//wp-content/uploads/2019/07/On-Birinci-Kalkinma-Plani.pdf, 1–2 (Last accessed October 15, 2019).
17. For further information see: United Nations, Sustainable Development Knowledge Platform, https://sustainabledevelopment.un.org/vnrs/ (Last accessed October 21, 2019).
18. Turkey's 2nd VNR, 2019.
19. United Nations, Sustainable Development Knowledge Platform, https://sustainabledevelopment.un.org/sdg7 (Last accessed October 15, 2019).
20. United Nations, Department of Economic and Social Affairs, Statistics Division, https://unstats.un.org/sdgs/report/2019/goal-07/ (Last accessed October 15, 2019).
21. Ibid.
22. Ibid.
23. United Nations, Sustainable Development Knowledge Platform, https://sustainabledevelopment.un.org/content/documents/195532018_background_notes_SDG_7Final1.pdf (Last accessed October 17, 2019).
24. Turkey's 2nd VNR, 77.
25. Ibid., 33.

26. Ibid., 77.
27. Turkey's 2nd VNR, 79.
28. "Purchase guarantee tariff is 7.3 USD cents/kWh for generation based on hydroelectricity and wind energies, 10.5 USD cents/kWh for generation based on geothermal energy and 13.3 USD cents/ kWh for generation based on biomass (including landfill gas) and solar energies." Turkey's 2nd VNR, 79.
29. Ibid., 78–79.
30. Turkey National Energy Efficiency Action Plan, 1.
31. Ibid.
32. Ibid., 4.
33. Turkey's 2nd VNR, 80.
34. Regulation on Renewable Energy Zones, "Yenilenebilir Enerji Kaynak Alanlari Yönetmeliği," http://www.resmigazete.gov.tr/eskiler/2016/10/20161009-1.htm (Last accessed October 21, 2019).
35. United Nations, Department of Economic and Social Affairs, Statistics Division, https://unstats.un.org/sdgs/report/2019/goal-13/ (Last accessed September 20, 2019).
36. United Nations, Economic and Social Council, "Special Edition: Progress towards the Sustainable Development Goals," https://undocs.org/E/2019/68, 18 (Last accessed October 2, 2019).
37. Ibid.
38. UNFCCC, 2018, "Turkey's 7th National Communication," 21.
39. United Nations, Sustainable Development Knowledge Platform, https://sustainabledevelopment.un.org/sdg13 (Last accessed October 12, 2019).
40. The World Bank, https://data.worldbank.org/country/turkey (Last accessed October 2, 2019).
41. UNFCCC, 2015, Republic of Turkey, Intended Nationally Determined Contribution, https://www4.unfccc.int/sites/submissions/INDC/Published%20Documents/Turkey/1/The_INDC_of_TURKEY_v.15.19.30.pdf, 2 (Last accessed September 20, 2019).
42. Turkey's 2nd VNR, 34.
43. Ibid., 110.
44. Ibid.
45. UNFCCC, 2018, "Turkey's 7th National Communication."
46. Ministry of Energy and Natural Resources (MENR), Energy Efficiency Strategy Document (2012–2023). https://www.resmigazete.gov.tr/eskiler/2012/02/20120225-7.htm (Last accessed October 25, 2019).

47. General Directorate of Energy Affairs, National Renewable Energy Action Plan for Turkey (NREAP), Yenilenebilir Enerji Eylem Planı (2014–2019), https://www.eigm.gov.tr/File/?path=ROOT%2f4%2fDocuments%2fEnerji%20Politikas%C4%B1%2fNational_Renewable_Energy_Action_For_Turkey.pdf (Last accessed October 21, 2019).
48. National Renewable Energy Action Plan for Turkey (NREAP), 19.
49. Turkey's 2nd VNR, 112.
50. UNFCCC, "Parties & Observers," https://unfccc.int/parties-observers (Last accessed October 18, 2019).
51. UNFCCC, "Republic of Turkey Intended Nationally Determined Contribution," 62–63.
52. Turkey's 2nd VNR, 112.
53. Ibid, 113.
54. UNFCCC, 2018, "Turkey's 7th National Communication," 19.
55. Ibid.
56. MENR, https://www.enerji.gov.tr/tr-TR/Sayfalar/Komur (Last accessed October 30, 2019).
57. MENR, https://www.enerji.gov.tr/tr-TR/Sayfalar/Elektrik (Last accessed October 30, 2019).
58. UNFCCC, 2016, "Report on Turkey's 6th National Communication," FCCC/IDR.6/TUR, http://unfccc.int/resource/docs/2016/idr/tur06.pdf, 29 (Last accessed September 30, 2019).
59. UNFCCC, 2018, "Turkey's 7th National Communication," 21.
60. Ibid., 50.

Bibliography

Legislative Documents

Disaster and Emergency Management Authority (AFAD) Strategic Plan (2013–2017), https://www.afad.gov.tr/kurumlar/afad.gov.tr/2402/files/afadsp2013-2017.pdf (Last accessed October 25, 2019).

General Directorate of Energy Affairs, Electrical Energy Market and Supply Security Strategy (2009), https://www.eigm.gov.tr/File/?path=ROOT%2f4%2fDocuments%2fEnerji%20Politikas%C4%B1%2fElectricity%20M

arket%20and%20Security%20of%20Supply%20Strategy%20Paper.pdf (Last accessed October 21, 2019).
General Directorate of Energy Affairs, Electrical Energy Sector Reform and Privatization Strategy (2004), https://www.eigm.gov.tr/File/?path=ROOT%2f4%2fDocuments%2fEnerji%20Politikas%C4%B1%2fElektrik%20Enerjisi%20Sekt%C3%B6r%C3%BC%20Reformu%20ve%20%C3%96zelle%C5%9Ftirme%20Strateji%20Belgesi.pdf (Last accessed October 21, 2019).
General Directorate of Energy Affairs, Turkey National Renewable Energy Action Plan (2014–2019), https://www.eigm.gov.tr/File/?path=ROOT%2f4%2fDocuments%2fEnerji%20Politikas%C4%B1%2fNational_Renewable_Energy_Action_For_Turkey.pdf (Last accessed October 21, 2019).
Ministry of Energy and Natural Resources (MENR), Enerji ve Tabii Kaynaklar Bakanlığı Stratejik Planı (2015–2019), https://enerji.gov.tr/File/?path=ROOT%2f1%2fDocuments%2fStratejik%20Plan%2fETKB%202015-2019%20Stratejik%20Plani.pdf (Last accessed October 21, 2019).
Ministry of Energy and Natural Resources (MENR), Enerji Verimliliği Strateji Belgesi (2012–2023), https://www.resmigazete.gov.tr/eskiler/2012/02/20120225-7.htm (Last accessed October 25, 2019).
Official Gazzette, Regulation on Renewable Energy Zones, "Yenilenebilir Enerji Kaynak Alanlari Yönetmeliği," http://www.resmigazete.gov.tr/eskiler/2016/10/20161009-1.htm (Last accessed October 21, 2019).
Presidency of Turkey, Presidency of Strategy and Budget, "Turkey's 8th National Development Plan," http://www.sbb.gov.tr/wp-content/uploads/2018/11/Eight-Five-Year-Development-Plan-2001-2005.pdf (Last accessed October 15, 2019).
Presidency of Turkey, Presidency of Strategy and Budget, "Turkey's 9th National Development Plan," http://www.sbb.gov.tr/wp-content/uploads/2018/11/Ninth_Development_Plan_2007-2013.pdf (Last accessed October 15, 2019).
Presidency of Turkey, Presidency of Strategy and Budget, "Turkey's 10th National Development Plan," http://www.sbb.gov.tr/wp-content/uploads/2018/11/Onuncu-Kalk%C4%B1nma-Plan%C4%B1-2014-2018.pdf (Last accessed October 15, 2019).
Presidency of Turkey, Presidency of Strategy and Budget, "Turkey's 11th National Development Plan," http://www.sbb.gov.tr//wp-content/uploads/2019/07/On-Birinci-Kalkinma-Plani.pdf (Last accessed October 15, 2019).
Presidency of Turkey, Presidency of Strategy and Budget, "Turkey's 7th National Development Plan," http://www.sbb.gov.tr/wp-content/uploads/2018/11/Yedinci-Be%C5%9F-Y%C4%B1ll%C4%B1k-Kalk%C4%

B1nma-Plan%C4%B1-1996-2000%E2%80%8B.pdf (Last accessed October 15, 2019).

Turkey's 2nd VNR, 2019, Sustainable Development Goals, "Strong Ground towards Common Goals," https://sustainabledevelopment.un.org/content/documents/23862Turkey_VNR_110719.pdf (Last accessed October 15, 2019).

Turkey's Action Plan to Combat Erosion (2013–2017), https://www.ogm.gov.tr/ekutuphane/Yayinlar/Erozyonla%20M%C3%BCcadele%20Eylem%20Plan%C4%B1.pdf (Last accessed October 25, 2019).

Turkey's Disaster Response Plan, Türkiye Afet Müdahale Planı (TAMP) (2013), https://www.afad.gov.tr/kurumlar/afad.gov.tr/2419/files/Afet_Mud_Pl_ResmiG_20122013.pdf (Last accessed October 25, 2019).

Turkey's National Climate Change Strategy, Türkiye İklim Değişikliği Stratejisi (2010–2023), https://www.gmka.gov.tr/dokumanlar/yayinlar/Turkiye-Iklim-Degisikligi-Stratejisi.pdf (Last accessed October 21, 2019).

Turkey's National Energy Efficiency Action Plan, Ulusal Enerji Verimliliği Eylem Planı (2017–2023), http://www.yegm.gov.tr/document/20180102M1_2018.pdf (Last accessed October 21, 2019).

Turkey's National Biological Diversity Strategy and Action Plan, Ulusal Biyolojik Çeşitlilik Stratejisi ve Eylem Planı (2007–2017), http://www.milliparklar.gov.tr/dosyalar/nbap.pdf (Last accessed October 25, 2019).

Turkey's National Drought Management Strategy and Action Plan, Ulusal Kuraklık Yönetimi Strateji Belgesi ve Eylem Planı (2017–2023), https://www.tarimorman.gov.tr/SYGM/Belgeler/Ulusal%20Kurakl%C4%B1k%20Y%C3%B6netimi%20Strateji%20Belgesi%20ve%20Eylem%20Plan%C4%B1/Ulusal%20Kurakl%C4%B1k%20Y%C3%B6netimi%20Strateji%20Belgesi%20ve%20Eylem%20Plan%C4%B1.pdf (Last accessed October 25, 2019).

Turkey's National Program and Action Plan to Decrease the Negative Effects of Climate Change on Health, İklim Değişikliğinin Sağlık Üzerine Olumsuz Etkilerinin Azaltılması UlusalProgramı ve Eylem Planı, https://hsgm.saglik.gov.tr/depo/birimler/cevre-sagligi/2-ced/iklim-degisikligi/Iklim_Degisikligi_Eylem_Plani.pdf (Last accessed October 25, 2019).

Turkey's National Strategy and Action Plan for Combatting Agricultural Drought, Tarımsal Kuraklıkla Mücadele Stratejisi ve Eylem Planı (2018–2022), https://istanbul.tarimorman.gov.tr/Belgeler/KutuMenu/KuraklikEylemPlani/Kurakl%C4%B1k%20Kitab%C4%B1%20%C4%B0c%20Sayfalar.pdf (Last accessed October 25, 2019).

Turkey's National Strategy and Action Plan to Combat Desertification, Çölleşme ile Mücadele Ulusal Stratejisi ve Eylem Planı (2015–2023), https://www.tarimorman.gov.tr/CEM/Belgeler/collesme%20belgeleri%20arsiv/Sayfa04/%C3%87%C3%96LLE%C5%9EME_16X23.pdf (Last accessed October 25, 2019).

Turkey's National Waste Management and Action Plan, Ulusal Atık Yönetimi ve Eylem Planı (2016–2023), https://webdosya.csb.gov.tr/db/cygm/haberler/ulusal_at-k_yonet-m–eylem_plan–20180328154824.pdf (Last accessed October 25, 2019).

Turkey's National Climate Change Adaptation Strategy and Action Plan, Türkiye'nin İklim Değişikliği Uyum Stratejisi ve Eylem Planı (2011–2023), https://webdosya.csb.gov.tr/db/iklim/editordosya/uyum_stratejisi_eylem_plani_TR.pdf (Last accessed October 21, 2019).

Turkey's National Earthquake Strategy and Action Plan, Ulusal Deprem Stratejisi ve Eylem Planı (2012–2023), https://www.resmigazete.gov.tr/eskiler/2011/08/20110818-13-1.pdf (Last accessed October 25, 2019).

Turkish Coal Corporation, TKİ Stratejik Planı (2015–2019), http://www.tki.gov.tr/depo/2017/TKi_StratejikPlan.pdf (Last accessed October 21, 2019).

Turkish Coal Corporation, TKİ Stratejik Planı (2019–2023), http://www.tki.gov.tr/depo/file/TK%C4%B0_2019_2023_Stratejik_Plan.pdf (Last accessed October 21, 2019).

Turkish Electricity Transmission Corporation, TEİAŞ Stratejik Planı (2015–2019), https://www.teias.gov.tr/tr-TR/stratejik-plan (Last accessed October 21, 2019).

Other Documents

Bertelsmann Stiftung and Sustainable Development Solutions Network (SDSN), Sustainable Development Report 2019, https://www.sdgindex.org/reports/sustainable-development-report-2019/ (Last accessed October 29, 2019).

Ministry of Energy and Natural Resources (MENR), https://www.enerji.gov.tr/tr-TR/Sayfalar/Elektrik (Last accessed October 30, 2019).

Ministry of Energy and Natural Resources (MENR), https://www.enerji.gov.tr/tr-TR/Sayfalar/Komur (Last accessed October 30, 2019).

The World Bank, https://data.worldbank.org/country/turkey (Last accessed October 2, 2019).

UNFCCC, "Parties & Observers," https://unfccc.int/parties-observers (Last accessed October 18, 2019).
UNFCCC, 2015, Republic of Turkey, Intended Nationally Determined Contribution, https://www4.unfccc.int/sites/submissions/INDC/Published%20Documents/Turkey/1/The_INDC_of_TURKEY_v.15.19.30.pdf, 2 (Last accessed September 20, 2019).
UNFCCC, 2016, "Report on Turkey's 6th National Communication," FCCC/IDR.6/TUR, http://unfccc.int/resource/docs/2016/idr/tur06.pdf, 29 (Last accessed September 30, 2019).
UNFCCC, 2018, "Turkey's 7th National Communication," https://unfccc.int/sites/default/files/resource/496715_Turkey-NC7-1-7th%20National%20Communication%20of%20Turkey.pdf (Last accessed September 18, 2019).
United Nations Development Programme (UNDP), "World Leaders Adopt Sustainable Development Goals," September 25, 2015, https://www.undp.org/content/undp/en/home/presscenter/pressreleases/2015/09/24/undp-welcomes-adoption-of-sustainable-development-goals-by-world-leaders.html (Last accessed August 2, 2019).
United Nations Statistics Division, Sustainable Development Goal indicators website, https://unstats.un.org/sdgs/files/Working-Group-ToR–Interlinkages.pdf (Last accessed September 18, 2019).
United Nations, Department of Economic and Social Affairs, Statistics Division, https://unstats.un.org/sdgs/report/2019/goal-13/ (Last accessed September 20, 2019).
United Nations, Department of Economic and Social Affairs, Statistics Division, https://unstats.un.org/sdgs/report/2019/goal-07/ (Last accessed October 15, 2019).
United Nations, Economic and Social Council, "Special Edition: Progress towards the Sustainable Development Goals," https://undocs.org/E/2019/68, 18 (Last accessed October 2, 2019).
United Nations, Sustainable Development Knowledge Platform, https://sustainabledevelopment.un.org/vnrs/ (Last accessed October 21, 2019).
United Nations, Sustainable Development Knowledge Platform, https://sustainabledevelopment.un.org/sdg7 (Last accessed October 15, 2019).
United Nations, Sustainable Development Knowledge Platform, https://sustainabledevelopment.un.org/content/documents/195532018_background_notes_SDG_7Final1.pdf (Last accessed October 17, 2019).
United Nations, Sustainable Development Knowledge Platform, https://sustainabledevelopment.un.org/sdg13 (Last accessed October 12, 2019).

7

To Grow or Not to Grow: Evolution of the Economic Paradigm as a Response to Climate Disruption

Małgorzata Zachara

Introduction

Climate and environmental hazards make the current phase of growth increasingly uneconomic—its costs overshadow its benefits. Furthermore, it is also becoming apparent that Western societies have already crossed the line marking the correlation between certain levels of income and an increase of subjective wellbeing (Frey 2008, 2018). On the one hand, economists emphasize the need to find policies for controlling climate change that optimize the trade-off between climate damages and lost opportunities for consumption or economic development; on the other, the scale of the environmental consequences of the present growth paradigm leads to the collapse of the major economic consensus linking economic growth per capita with human well-being and social cohesion. As environmental issues, climate change prevention and sustainable

M. Zachara (✉)
Faculty of International and Political Studies,
Jagiellonian University, Krakow, Poland
e-mail: malgorzata.zachara@uj.edu.pl

E. V. Shabliy et al. (eds.), *Discourses on Sustainability*,
https://doi.org/10.1007/978-3-030-53121-8_7

growth have become basic elements of the 'human happiness' concept, a 'green new deal' which places greater emphasis on climate mitigation and contentment, rather than GDP, is now seen as a strategy for the future (FitzRoy et al. 2012).

This chapter presents the major aspects of the academic debate about the philosophical questions connected to the 'good life' principle that is being incorporated into economic policies and reflected in market practices.

There is no doubt that climate change is an economic issue. Climate disruptions directly interact with all human social structures, but as environmental degradation is largely attributed to economic growth, economic activities stand at the centre of the complex phenomena. Environmental shifts and their harmful effects produce challenges in other systems, leading to the introduction of new social models based on coping strategies and preventive measures. These tasks, as aspects of policy-making processes, require the transformation of established notions of prosperity, wealth and economic development. So climate change has brought not only a need for a redefinition of the operational schedules of economic systems, but also touches the issues of Western self-identification and questions the historical foundations of modernity. Rather than celebrate the achievements of modernity, the advances of science, and industrial mastery over nature, environmental discourses shed a different light on the Western road to economic and social primacy. The quest to conquer nature in order to rid humanity of the pernicious threat of scarcity whether in terms of food, shelter or basic needs, and to fight disease, stood at the centre of the project of civilizational progress. A desire to improve the material wellbeing of the people was in the background of the revolutions in thinking about markets and the development of social, political and economic organization in the eighteeenth and nineteenth centuries. This task couldn't have been completed without a mass mobilization of energy—the engine of the rising industrial age. Beginning with the industrial revolution and technological acceleration thereafter, the human consumption of fossil fuels significantly raised carbon dioxide (CO_2) levels, but also enabled the development of a sophisticated social organization and further technological development, and brought mass prosperity to the countries that embraced the principles of progress. But this developmental moment in world history is coming to an end, with the emergence of rising concerns connected to environmental risks—climate change being the most critical. The reliance on automotive transportation, rising resource and energy demands, the destruction of natural habitats and the consequences of pollution are seen as inevitable consequences of capitalistic overreach. There is a growing recognition of the fact that infinite growth is impossible in a finite world. The process of the social construction of environmental risks and climate change that has been underway for about 30 years now has created the entry conditions for policy planning and political activity. This process has taken the form of discursive activities that demonstrate that the ways in which climate change is described, discussed and understood influences to a great extent the process of economic and social paradigm change. In socio-economic terms, the scale of the environmental hazards redefined the sense of modernity, founded on the ideal of economic growth, seen as a mechanism for the creation of social wealth and human emancipation. On the one hand, economists emphasize the need to find policies for controlling climate change that can optimize the trade-off between climate damages and lost opportunities for consumption or economic development; on the other, the scale of the environmental consequences of the present growth paradigm leads to the collapse of the major economic consensus linking economic growth per capita with human wellbeing and social cohesion. Climate and environmental hazards make the current phase of growth increasingly uneconomic—its costs overshadow its benefits. Furthermore, it is also becoming apparent that Western societies have already crossed the line marking the correlation between certain levels of income and an increase of subjective wellbeing (Frey 2008, 2018). As environmental issues, climate change prevention and sustainable growth have become basic elements of the 'human happiness' concept, a 'green new deal' which places greater emphasis on climate change mitigation and contentment, rather than GDP, is now widely seen as a strategy for the future (FitzRoy et al. 2012). The concepts of 'sustainability', 'green growth' and 'degrowth' occupy central positions in the public debate on the relationship between economic growth and the environment, but this equation has a third element behind it: social welfare. Analysis of this part of the debate reveals that economic growth taking a central position in the environmental destruction as well as in environmental care is not an objective *per se*, but rather a means to achieve certain ends. This chapter presents the major aspects of the academic debate about the philosophical questions connected to the 'good life' principle that is being incorporated into economic policies and reflected in market practices all around the world. At least three complex levels of analysis can be traced in this context:

The individual dimension of climate change policies—while climate disruption is most often portrayed as a challenge for the international community, states or governments, the essence of the social change in the environmental sphere lies within the beliefs, ethical orientations and behaviours of individuals. The question of whether people are ready to give up the energy-intensive lifestyle, rewrite the principles of the traditional economy and agree on sacrifices in order to follow the rules of the more sustainable models, is a decisive factor in building the adaptive capacity of the Western societies. Furthermore, people in industrialized nations today can determine the fate of people in developing countries and that of future generations everywhere. Patterns of individual decision-making are increasingly important given the existence of polycentric climate governance forums and bottom-up initiatives, in which diverse actors seek to realize diverse benefits or avoid costs associated with environmental hazards (Ostrom 2009).

The tragedy of the commons—the earth is inherently a global natural resource which, in terms of its use and governance, represents for many analysts a model case of a 'tragedy of the commons'—the public good dilemma made famous by Garrett Hardin (1968). The 'tragedy' refers to the inability of groups to manage common resources: individuals and groups in pursuit of their self-interest (or what Hardin calls 'the sum of separate ego-serving decisions') (1998: 682), overexploit the planet's atmosphere. As a result, the unmanaged commons could easily be ruined. The tragedy of the commons analogy reveals the clashing logics of the liberal economic order based on self-interest and nature as a global public goods imposing common interest rationality.

The collective action problem—at the centre of collective climate change action stands a mechanism according to which the parties are likely to act in a short-term and self-centred manner, because while the costs of acting are borne individually the benefits of acting are shared by all. In this situation individuals, states or small groups of them will neither be able to substantially reduce local harm, nor fully internalize the benefits of their investments. Such a deadlock has been partially broken by raising climate risks awareness and influencing the preferences of the Western electorates or by seeing sustainable solutions as an engine for economic growth. There leaves, however, the still-unresolved question of reaching a sense of shared interest and trust that would secure the engagement of developing-world stakeholders.

The Collapse of Consensus

With rising environmental concerns and politically contested approaches emerging to remedy them, a largely theoretical debate explores the possible linkages between the economic model and the environment. While natural resources and geographical and climate conditions have always been powerful factors in the building of a successful economic system, human relations with the environment have become so complex that they are transforming the boundaries of economic thinking. Adam Smith's seminal work on the Western market economy—*An Inquiry into the Nature and Causes of the Wealth of Nations*—argues that the wealth of nations is founded on a base of natural conditions (with access to the sea and a moderate climate playing the major roles). These, combined, with free-market principles and loose governmental control, provide the basis for wealth and prosperity. The open market economic model, promoting international exchange over protectionism, individual entrepreneurship over monopolies, and superseding money as a source of wealth marked the capitalistic economic paradigm for more than two centuries. From the late eighteenth century Western economists and policy-makers equated material prosperity with the advance of civilization. Modern societies and their lifestyles were founded on their distance from nature, by providing civilizational capsules independent of environmental conditions. Air-conditioned and centrally heated city apartments and offices have become the symbol of post-environmental modernity, a condition in which man ceased to be at the mercy of natural conditions, protected by concrete walls and the secure protective box represented by his car. Successful economies demonstrated their powers by dramatically transforming natural landscapes and creating lifestyles based on material wealth and a belief in individualism. The triumph of civilization in the Western world brought about not only a spectacular improvement in living conditions for millions of people but revealed a fundamental tension, between the natural world and the human hunger for growth. Growing imbalances between the purposes of market expansion, combined with the requirements of the consumerist societies and frameworks of development imperatives, have led to major reorientations within the area of economic principles. In the second half of the twentieth century, it became evident that there is a link between the capitalistic ideal of prosperity and environmental degradation. The leading economic model, originating with Smith's ideas, is premised on ever-expanding growth, yet the real world system in which it operates is restricted by finite resources and a finite capacity to absorb waste.

The long historical period of the successful development of capitalism and neoliberalism was partially founded on the belief that free-market forces always respond with an adequate solution to societal and environmental problems; the human and environmental costs of the exponential growth, for example, will be mitigated by new technologies or policies that will enable the economic model to survive. It seems, though, that the climate change crisis revealed the failure of the technological promise, putting the established model of growth and environmental wellbeing on the opposite side of the spectrum. For many economists, commentators and policy-makers, we simply cannot have them both. The ecological sustainability of human societies is increasingly in question, as the scale of many environmental problems continues to escalate, with global climate change being one of the most pressing. The essence of this issue lies not only in arguments about the need to adjust the existing economic practices so as to reflect higher environmental standards, but more in a redefinition of the logic of growth. The international debate has reached the point at which it has been suggested that it is beyond doubt that the impact of climate change may be potentially catastrophic for the integrity of social and ecological systems around the world (Parry et al. 2007).

This risk has for a long time been affecting the way people and states manage their economic affairs, but it is increasingly recognized that the challenge goes beyond making an adaptation to new climate conditions and implementing procedures to mitigate greenhouse gases. William D. Nordhaus, a 2018 Nobel Prize Laureate in Economics, diagnosed the situation in straightforward terms: 'Today, we have a climate problem because markets fail and fail badly in the energy sector. They fail because of one of the by-products of the use of energy, in particular fossil fuels, is unpriced' (Nordhaus 2018). The Stern review considers climate change to be a '…market failure on the greatest scale the world has seen' (Stern 2007, 27).

Capitalism has been placed at the heart of the challenge of confronting climate change, and any serious attempt to address this at the level of the global climate must interfere with the established structures of globalization. The question is, however, whether the correction of the market's failures that cause or exacerbate climate change is possible within the established framework, or needs to go beyond it. Political reactions to the environmental consequences of the established patterns of growth have long been embedded in the nature of political decision-making patterns, oriented towards short terms gains and the securing of electoral rewards.

The magnitude of the destruction revealed over time, has further undermined the traditional model of growth (Gore 1992; Brown 2006; Vanderheiden 2008), resulting in an increased interest in a 'green' or 'natural' capitalism (OECD 2011). All such concepts break with the reduction of nature to the status of simply being a raw material (natural resource), as has long been common in the classic economic views. Sustainability becomes the most powerful label for environmental concern in relation to traditional market practices. The central tenet of the older idea of growth is the maximization of economic opportunities rather than the challenges to environmental sustainability arising from the market. Even if it is green, growth still stands at the centre of this view—the aims of economic activity are, on the one hand, to protect the environment, while on the other to allow for economic growth. Especially in the period following the Great Recession of 2007–2008, the idea becomes attractive for policy-makers in the Western world, as the demand for a policy-driven stimulation of demand has become stringent. Nevertheless, the legitimacy of the current model of turbo-capitalism has been seriously and simultaneously shaken by the scale of the global recession that began in 2007 and fiasco of the Washington consensus' mechanism guiding the global financial governance institutions in their policies for supporting developing economies. These powerful agents in advancing economic discourses and shaping the reality of the global system have highlighted the risks of relying too much on market-led globalization. In the second decade of the twenty-first century it became clear that the existing model just does not stand up to scrutiny and that economic growth in the West entered a period of stagnation, largely because the global financial system has long since ceased to serve as a tool of development for the material economy and has become a goal in itself. As a result, an increased interest in 'Green Keynesianism' has arisen, defining the crisis of neoliberalism in terms of a new opening for a more economically active state to re-assert itself in fostering industrial, social and technological advance for the public good. The global financial crisis ended the Western dominance in the management of global economic affairs, while 'the rise of the rest', which has strengthened the position of the emerging markets, introduced even greater competition of ideas and models.

It has been widely agreed that environmental problems are a manifestation of the market failures and that public intervention has become a common solution. Governments embraced the sustainable approaches that dominated the Paris climate summit and the post-2015 sustainable development goals (SDGs). But reality proves that the wellbeing deriving from a preserved environment cannot be easily attained through market transactions, even if these are supported by states. The measures that have been taken throughout the last 30 years to deliver a harmonization of economic growth, social welfare and environmental protection are increasingly seen as ineffective. Worsening ecological and economic crises and continuing social deprivation brought a new consensus that an entirely new paradigm is needed to stabilize the relation between nature and human economic affairs. However, for the time being, there is no clear objective that could replace the established and grounded notion of exponential growth.

The array of proposals for a new model of growth vary from different versions of sustainable growth (Bäckstrand and Kronsell 2015) through the steady-state economy (Blauwhof 2012), ecological citizenship (Dobson 2003) and the green economy (Green Economy Coalition 2012) to degrowth (D'Alisa et al. 2014; Jackson 2009), but all of them are united by the idea of linking social with economic goals and meeting the needs of both the present and future generations of humankind.

Degrowth poses the question of the limits of the physical and social environment confronted with the amount of consumption in the global, capitalist system. It represents the most radical optics, directly placing capitalism opposite of social and physical survival (D'Alisa et al. 2014). According to this view sustainable development cannot provide a solution to our current ecological problems and the imbalances within the economic system, because it is rooted in the growth paradigm. The obsession with growth brought the world to the deep environmental and social crisis, so the ways out are to be sought outside the capitalistic framework. While the intellectual foundations of degrowth have confidently taken on the role of opposition to a firmly rooted ideal of economic growth, the concept hasn't been translated into practical terms and institutional bases. It is rather the ideological formulae standing in opposition to 'sustainability' or 'green economy', programmes of action have been used to address the policy challenges of the environmental and economic crises. Interestingly, there are also voices that identify the introduction of the green economy as a necessary phase of market reorientation in the search for an optimal version of the system (Friedman 2008). This narrative, originating in Schumpeter's notion of 'creative destruction' (Schumpeter 1942), sees the climate crisis as an opportunity to destroy over-accumulated and inefficient capital, by reducing overcapacity and creating openings for new market players. In this vision, the global economic system must be transformed towards the new low-carbon energy regime able to mitigate greenhouse gas emissions (Mabey

2007), which will create a new engine of capitalist growth and accumulation in the form of 'green technologies' (Bina 2013). According to this view, which has been drawn into the forefront of environmental discourse and policy debates at the national and international levels (Ferguson 2015), the mechanism will remain the same, but the negative consequences of economic transactions will be eliminated. On the practical level 'the global Green New Deal', 'green stimulus' and 'green growth' are oriented towards investing 1.5–2 per cent of global GDP per year in raising energy-efficiency standards and expanding the clean and renewable energy supply in order to reduce global CO_2 emissions by 40 per cent relative to today within 20 years, while also supporting rising living standards and expanding job opportunities. To reach these goals the consumption of oil, coal and natural gas will also need to fall by about 35 per cent over this same twenty-year period—an average rate of decline of 2.2 per cent per year (Pollin 2019). So the dominant policy orientation is based on the 'fixing' instead of 'shifting' of dominant socio-economic paradigms (Bina 2013). In fact, green technologies are suggested to be the foundation of a new technological revolution that will secure long-term growth and mitigate the social side-effects of automation.

Such a scenario, proposed by global governance institutions such as the World Bank or IMF and policy-makers in the most developed countries, tries to handle the economic efficiency crisis but does not take into account the 'legitimacy crisis'. The established and accepted model of economic growth has raised millions of people out of poverty, opened up markets, and spread technological achievements and educational opportunities all around the world, while at the same time producing major imbalances. 'Economic growth' per se cannot be seen as an engine of social development, as it has been proven that it makes no reference to the distribution of the benefits of growth.

Despite the general sense of urgency created by the media and increasing awareness of the importance of climate issues, the conceptual picture of the green or post-growth economy is highly fragmented. The issue of sustainability and green modes of economic development is even more divisive in the light of the most serious problem of the contemporary phase of global capitalism—inequality. Throughout the two centuries of the application of neoclassical economic principles, the gap between rich and poor continued to widen: on a per capita income basis, the rich-to-poor ratio was 2:1 in 1800, 20:1 in 1945 and 40:1 by 1975. The richest 20 per cent of the world's population accounts for 82.7 per cent of global income, while the poorest 20 per cent earn 1.6 per cent of global income (Waters 1995; OECD 2015). Inequality has come to the fore as a core concern across the world as a major manifestation of the assumption that the dominant economic model is corrupted, as it undermines social solidarity and causes deprivation and social exclusion.

Even if a consensus on the relationship between the status or role of the environment and economic activity had been reached, there is a striking disconnect between the normative visions derived from climate change discourses and the actual organization of economic activity. The attempt to adapt basic market principles to the newly defined concepts of well-being remains a challenging puzzle, illustrated by three dilemmas structurally embedded in the market logic:

The value dilemma: it has been recognized that dysfunctions connected to the capitalist system stem from a misconception of economic value. GDP as a measure of national wealth has long been questioned, as not reflecting the overall quality of life and distorting the value of socially important activities such as care work or public expenditure. The metrics do not include externalities; assets and activities in the non-monetary and informal economies are ignored. A number of alternative measurements such as the Index of Sustainable Economic Welfare and the Genuine Progress Indicator have been devised to address this imbalance, but the relation between the human dimension of market activity and market value still hasn't been clearly explained. There is a growing need to create a new definition of value, which would refer to the economy rather as a general category of human behaviour and relations, and not only as a profit-generating system.

The quality of life dilemma: the market, apart from politics, is one of the social mechanisms responsible for the generation and distribution of human wellbeing (Lindblom 1977). Research on how happiness relates to material wealth clearly documents that people report higher levels of subjective wellbeing if they live in wealthy rather than poor nations.

However, once the basic needs of individuals are secured, material affluence does relatively little to improve happiness. Market narratives of the 'good life' are clearly connected with the attainment of material success, but they do not mention the costs associated with this culturally acclaimed path—the economic insecurity, personal loss of autonomy, inequality or status anxiety (Lane 1978). With the rise of ecological awareness, visions of what constitutes 'the good life' are coming to focus around relations with nature and with others.

The financialization of the economy dilemma: from the earliest days of human culture, economic activities were focused on the improvement of the quality of life and protection against unfavourable twists of fate. In the aftermath of the industrial revolution, however, economic principles shifted towards the interests of the major market actors and maximization of shareholder value. During the last decades of the twentieth century, globalization created the main framework of the economic system, dramatically increasing the influence of the financial markets. This marked the beginning of a phase in capitalist development in which profits accumulated increasingly through financial channels, weakening the connections between the financial services industries and the requirements of the real economy. In effect, the economic system primarily secures short-term economic growth and corporate profit maximization but does not necessarily support objectives pursued by society as a whole, and at times jeopardizes the quality of life prospects of whole social groups.

The Climate Divide

Climate change has long been presented as the major manifestation of the realness of globalization and an argument supporting the necessity of rejecting local definitions of political interests or social conditions: 'The power of this new story of climate change lay in its global reach. It was the global temperature and sea-level that were rising, it was the global climate system that was changing, it was global climate models that were telling us so. It was not merely climate change; it was global warming' (Hulme 2010). The idea of the climate transformation has been constructed as bearing a distinctive universalising, planetary sense, but the reactions and measures meant to control the speed of destructive environmental consequences have to be local, and undertaken by states, cities and individuals. Here a distinction can be drawn between rich and poor on two levels: highly developed and developing countries, as well as rich individuals whose lifestyles contribute the most to the climate change damage and poor people, who are likely to suffer the most (Harris 2010, 130). So climate change as a divisive factor is to be found both within states, and between states, and rises to the level of generations, as the consequences of our present actions will be felt in the future.

Accountability is the essence of this debate: 'When people in an American city turn on the air-conditioning or people in Europe drive their cars, their actions have consequences. Those consequences link them to rural communities in Bangladesh, farmers in Ethiopia, and slum dwellers in Haiti' (UNDP 2007). This local dimension of the climate change experience, and the fears and needs associated with it, has been translated into climate policies in local political and social agendas. In this way, climate change has become part of the new imperialism debate in terms of the South–North divide, as well as in peace and development studies. All these areas make reference to these questions: who is responsible, who should bear the costs, which interests will be impeded by agreed strategies, to what extent should such severe consequences of climate change as rising sea levels be mitigated by collaborative action? Climate change negotiations mark an interesting moment in the development of the sense of globalness, because in trying to deal with these challenges a collaborative approach is essential.

South–North division: different economies exhibit different degrees of complexity, which determine their vulnerability to climate change impacts. Most developed countries in the Western world have complex economies that offer many sources of income and are more resilient during periods of stress, and on the whole, they are not vulnerable to extreme weather conditions. The opposite situation is to be found in the low-income countries where extreme events such as hurricanes can cost more than 10 per cent of the gross domestic product (GDP) (LinnerothBayer et al. 2005). Climate change is and will most negatively affect those living in less developed countries, who have done the least to contribute to the causes of climate change, while those living in developed countries, who have contributed the most emissions, will likely suffer the least. It is estimated that developing countries 'would bear some 75 to 80 percent of the costs of damages caused by the changing climate' (World Bank 2009: xx). Geographic locations, low income and low institutional capacity put developing countries in unfavourable positions on the map of climate change consequences. Because of this, only a small number of developing countries can afford to introduce mitigating strategies. Their greater reliance on climate-sensitive sectors like agriculture (Nanda 2009) makes them particularly vulnerable. These countries, with their limited energy use, are facing grave risks, although they have contributed little to climate change. For example, the Solomon Islands has already lost five small islands as a result of climate change and yet its emission rates are among the lowest in the world. Such asymmetry between responsibilities and impacts creates a 'double injustice' narrative on the side of the global South. Such asymmetry between responsibilities and impacts creates a 'double injustice' narrative on the side of the global South. As a consequence, a longstanding economic global divide is further exacerbated by climate change, which is growing to be a principal source of tension between developed and developing countries.

Right to development: the risks generated by climate change are inherently global and require global responses. What stands as the main obstacle in the process of designing them is the divergent logic of the climate discourses in the South and in the North. The Western world applies the narrative of the protection of nature, while developing countries make human wellbeing the primary point of reference. This means that during global climate negotiations, developing countries have been consistently stressing their 'right to development' and the unfairness of having to carry the same economic burden as highly industrialized societies. In fact, as the mitigation measures are implemented only in Western countries, their effects expose developing societies to greater risks. The model example is provided by the practice of converting land to biofuel production, which in recent decades has driven up the prices of essential foodstuffs and thus cut the real incomes of poorer peoples. Within the debate on the right to development, seen as a core principle of the concept of climate justice, it has been recognized that climate disruption is likely to set back human development in some states that are making progress. While addressing this problem, global institutions have to deal not only with the challenges of low-carbon development, but multidisciplinary issues of social, gender equality, economic, political and cultural concerns.

Climate conflicts: climate change-driven extreme weather and related disasters can damage economies, lower farming, and livestock production and intensify inequality among social groups. These factors, when combined with other drivers of conflict, may increase the risk of armed conflict. So, while there is no direct relation between climate change and conflict, the assumptions of its influence on generating and exacerbating tensions in the world are widely shared (Stern 2007; Burke et al. 2009). Stockholm Peace Research Institute study (2016) on the relationship between climate-related change and local, violent conflicts in East Africa and Sudan confirms that conflicts around natural resources—land, pasture, water—are linked to the worsening conditions for their provision due to climate disruption. Such a link is identified in other empirical studies, but the conclusions of these are not consistent as the climate change narrative is sometimes used as a misleading explanation for conflict to cover political and economic variabilities and structural conditions that are likely to shape the security situation in the vulnerable regions. Nevertheless, it is becoming increasingly realized that the reverse relationship is much more certain: societies already in conflict are particularly exposed and ill-equipped to meet the impacts of climate change. This division is shaped along developmental lines: places with widespread poverty, ineffective governments and direct dependence on natural resources face the gravest risks from the changing climate.

Ethics and Imaginaries

Foucault's (1980) analysis of power reveals how social practices constitute the boundaries of discourse, determining 'what is and what is not, what can be done and what cannot, what should be and what should not' (Clegg 1989). In the constriction of climate change imaginaries, enduring visions provide orientation towards these highly complex issues. Ideas of sustainability and climate risks applied to Western imaginaries of the economic practice produced new sensibilities, new knowledge and a new understanding of social relations (Norton 2002). This catalogue comprises narratives that point towards transcendent norms, and promote a moral backing for the desired change in economic patterns. In that sense, a climate change action plan is a cultural production, based on normative visions of the economy that do not stand in opposition to a concern for the environment. These visions are reflected in the political concepts and policies described above, but are also transmitted in the form of virtues, in the Aristotelian sense, as elements of a highly personal programme for the individual. Two major sources of the new ethics can be identified: environmental and social limits to growth. These two categories are organized around the same logic inevitable in market practices. Economic growth, spreading consumption and population growth negatively affect environmental resources like the ozone layer, drinkable water or wild nature, making them scarce, just as the marketization of social practices makes deep and meaningful relations, a sense of belonging or community values scarce. These are two sides of the same coin of the shortcomings of the traditional economic philosophies and practices, identified and analysed as early as in the 1950s, which found their practical manifestations in rising environmental awareness and changing social norms only recently. Long-acclaimed cultural features of the modernistic culture—consumerism, individualism and capitalism—clash with the requirements of the climate crisis and put the beneficiaries of the world of prosperity into the trap of anxiety and alienation which as Hannah Arendt wrote 'become an everyday experience of the ever-growing masses of our century' (Arendt 1951: 78). The change was brought about by the scale and the paste of the economic change and the realization that this transformation will be even deeper in the future with the growth of industry 4.0, from robotics and the Internet of Things to AI and cognitive computing. The nature of work has been altered, work security has been severely diminished and the established model of individual prosperity-building is failing. In many parts of the Western world younger generations face the prospect of not only being unable to surpass the standards of economic achievement of their parents' generation, but even to reach them. The sense of disappointment with the traditional concepts of growth and the magnitude of these transformations create an incentive for change. As a result, economic orientations towards growth have largely ceased to be seen as a rational choice for humanity, and new elements conditioning human behaviours, not covered by classic rational choice thinking, such as empathy or integrity, come into play in economic perspectives. The relationship between consumerism and the quality of life—previously approached only hesitantly by economists, is currently part of the mainstream debate. The traditional view, that consumption must be yielding welfare and that otherwise it wouldn't be occurring, is being challenged on theoretical (Sen 1977; Veblen 1909) as well as practical grounds.

Members of the most recent generations (Millennials, Generation Z) are the first to fully experience the imbalances and risks associated with the imperative of unbridled economic growth and are characterized by a deep generational awareness of the climate transition. Lifestyle-related demographic shifts, including decreased employment, falling accessibility of student loans and job market disruption factors foster a reconfiguration of the current principles. This heavily promoted change in individual behaviours produced forms of resistance that address the whole system of socio-economic interactions not consistent with the ideal of environment-friendly ways of life. This resistance applies to consumerism, producing as it does waste and social exclusion, economic arrangements that cannot protect people from material fragility in daily life, cultural patterns based on mercantilism, and neoliberal definitions of success. Within the framework of post-structural theory, resistance becomes a potential site of transformation, 'the means through which individuals change social processes and structures and build alternatives' (Sage 2007: 4707). This process has already started influencing young people's understandings of identity and the origins of Western economic thought. Millennials, born between 1980 and 1996, are the first Western generation to be environmentally conscious from birth and, additionally, they grew up in a period of economic stagnation. Research indicates that the majority of Millennials would pay more for environmentally friendly services, products or brands (The Nielsen Company 2015) and that they expect a more environmentally sensitive workplace (Guevarra 2010).

Members of Generation Z, born between 1995 and 2012, are aware that they will bear the brunt of climate change impacts and will largely be responsible for finding ways of adapting to them (Taylor and Keeter 2010). They are seen as much less materialistic than previous generations, radically inclusive and likely to make access rather of ownership a new form of consumption. Integrity is increasingly important in market relations—70 per cent of Gen Zers say they try to purchase products from companies they consider ethical (McKinsey 2018). The discourse on the moral character of global climate change is indeed reflected in individualized lifestyle choices that take the form of 'governed responsibilization' (Norgaard 2011; Ojala 2005). Two principal dimensions of responsibility in relation to environmental risk are connected with agency and obligation. Attributing responsibility entails identifying an actor as an agent, who identifies a normative sense of responsibility as a duty (Bickerstaff and Walker 2002). Individuals are assigned responsibility for the state of the environment and are held accountable, judged and sanctioned through this lens. In the everyday practice of avoiding environmentally hazardous behaviours, citizens are instructed to become prudent subjects who must 'practice individual responsibility' by informing themselves, instructing others and adjusting their lifestyles to the growing dangers of climate change. The environmental concern comes to be seen as a common set of values that are self-evident and immune to deep questioning or criticism.

Such a public construction of responsibility entails an individual sense of self-efficacy that is shaped by perceptions of the responsibilities of the state and of other institutional actors. The trend can also be seen, though, as an expression of greater autonomy of individuals or as a manifestation of the fact that contemporary systems of governance are ill-equipped for the task of facing complex problems like climate change. This view is reflected in the influential 'risk society' hypothesis proposed by Urlich Beck (1992) at the early stage of global economic acceleration and integration. Climate change and environmental hazards are among the manifestations of unintended and unforeseen side-effects of modern life, that backfire on modernity and the identity of the Western world. Ecological crises, for example emerge from the success of industrialisation and the ubiquitous spread of the city as a primary gathering point for commerce, work and living space. So the new risks constitute new forms of social organization and social imaginaries, and these differ fundamentally from that of the pursuit of wealth, around which society has to a large degree previously organized itself (Giddens 2006). Industrial society was a society of scarcity, focused on the production of wealth and primarily organized around generating economic growth. Environmental harms were made possible, because the logic of wealth production dominated the logic of risk generation. Postmodern risk society is an affluent society, defined not by scarcity but by risk—here the relation is reversed. Paradoxically, within such a social model the notion of responsibility itself becomes problematic: '…risk societies are characterized by the paradox of more and more environmental degradation perceived and possible coupled with an expansion of environmental law and regulation. Yet at the same time, no individual or institution seems to be held specifically accountable for anything' (Beck 1999: 149). A state of 'organised irresponsibility' becomes a norm: an increasing variety of environmental and technological risks are represented and perceived in contemporary culture as presenting a significant threat and often as being associated with failures of responsibility and control.

The overwhelming success of the new market trends like the sharing economy, crowdfunding, co-working, the reputation economy and the experience economy may also serve as an illustration of the Western social model's evolution, backed by a transformation of the understanding of prosperity and well-being. These all represent different market principles—the sharing of commodities instead of ownership, and consumer cooperation instead of producer competition. Acquiring, providing or sharing access to goods and services are facilitated by community-based online platforms, and take the form of peer to peer (P2P) activities (Taeihagh 2017). Social networks supported by technology have become powerful resources used to meet the needs of millions of users, beyond the established market rules, making the P2P economy one of the fastest-growing business trends in history. PricewaterhouseCoopers estimates that by 2025 global revenues from sharing in just five sectors (travel, car sharing, finance, staffing and music and video streaming) will have increased from \$15 billion in 2015 to \$335 billion (PricewaterhouseCoopers 2015). The idea of collaboration is embedded in the mechanisms of trust, recognition and collaborative finance, empowering users that gain access to the tools through which the market relation can be designed and built. 'Interactions', 'connectedness' and 'similar interest' all belong to the sphere of social capital, built through the channels of the new markets. These trends are generally seen as more individualistic—in the sense that they better respond to personalized needs and are anti-systemic—taking place beyond the structures of the traditional economic system, and more democratic, as the users enjoy increased decision-making powers. Sharing underused goods is also considered a contribution towards environmental sustainability and building value on the basis of a different set of motivations, rather than purely financial gain. One of the strongest incentives for participating in crowdfunding is the desire to belong to a social group created around innovation or idea and a sense of altruism (Gerber and Hui 2013). Social networking or crowdfunding platforms serve as vehicles for proactivity and answer the needs of individuals for competence, autonomy and relatedness (Ryan and Deci 2000). As McLaren and Agyeman note (2015: 64), 'The sharing economy has sparked a forest fire of excitement in terms of its potential to variously change the way we do business, empower previously powerless people, save resources, and increase our social closeness or civicness'.

The 'Cause Economy' has also become a powerful trend associated with networking and collective action, gathering people that share the notion of becoming involved and participating in the creation of something new. The real value for participants is not connected with financial rewards, thereby forming one of the major disruptive features of these trends, reflecting the growing need to reconfigure the ethical basis of market practices. The sharing-based economy provides the space in which reputation might develop into exchange value, translating effect into a form of market gain (Arvidsson and Peitersen 2013; Gandini 2016). As a consequence, the trends based on the 'sharing' or 'reputation' economy have been widely envisaged as a new *socio*-economic model supporting collaboration, access to, and the socialization of value production. There are voices within the debate warning that the 'sharing economy' is not actually about sharing at all (Eckhardt and Bardhi 2015), but the scale of these practices clearly indicates an evolutionary direction of the relationship between society and the market.

The neoclassical economic theory of welfare is based on the notion of a social welfare function supposed to capture all conceivable social aspirations. Neoclassical economic assumptions of 'individual optimisation' and 'utility maximization' define *homo* as *oeconomicus*. This hypothesis is starting to be overshadowed by research indicating that intrinsic motivation and immaterial rewards trump cash as dominant motivators to participation in different market practices (Hemer 2011; Harms 2007). Human beings are a highly social species, so their decisions are influenced by the decisions of others—social norms and roles play critical roles for individual behaviour, and create a base for individual rationality construction (Akerlof and Kranton 2000). Economic anthropology demonstrates that economic behaviour is itself politically and culturally constituted. Elements of social and cultural capital, cooperation, community and trust have always played a role in value creation, but it seems that they are becoming increasingly and directly translated into economic capital. All the examples of market practices discussed here are rooted in the shared meaning brought about by imaginaries defining the human relation to nature. Those related to climate change are the most powerful, not only because of their sense of urgency, but also the way in which they interact and co-evolve with economic, policy and technological aspects of energy systems to constitute coherent value regimes and a widely shared sense of legitimacy.

Conclusion

The reality of climate change is becoming central to the evolution of the economic model in the Western world, increasingly framing understandings of other developmental relations and national agency within the environmental area. Economic and ecological crises brought about the new conceptualization of the human-nature relation, which resulted in the creation of new public imaginaries with regard to individual and group activities, founded on a different value system. The system of meaning represented in these discursive and imaginative spaces became highly influential and marked the end of the ascendancy of neoclassical theory. The omnipresence of risk, fear and an increasing distrust of science and technology and its profit-driven outcomes, marks the downturn of the Enlightenment idea of rational progress. While growth has traditionally been associated with civilizational development, the accumulation of wealth with success, and economic strength with political power, progress is increasingly defined in more technological and human-oriented terms, and in some aspects appears to be rather post-materialistic. As a market actor, the consumer is no longer seen as insatiable, or independent, as one whose economic preferences are unaffected by those of others. All the programmes of transformation, from sustainability to degrowth, see *homo oeconomicus* in a new context, as more sensible of the common good, responsible, and likely to include the external consequences of their actions in their economic calculations.

The universal horizon is also an important aspect of the debate, as climate change creates global dangers that are unevenly distributed around the world. Strategies for fighting climate change, directly and indirectly, intersect with economic development, as Western countries are facing redefinitions of economic practice or abandoning the growth principle, while developing countries are expected to introduce preventive measures that directly impact their perspectives for growth. The consequences of climate change for security, human health and prosperity are, then, asymmetrical primarily due to the fact that resources and wealth are distributed unevenly. That which presents the fundamental challenge for the well-developed world seems deeply unfair in the developing countries, given the fact that economic growth is the most powerful instrument for reducing poverty and improving the quality of life in these societies.

Although a coherent model of adaptation to the new environmental challenges has not yet been established, the debate itself leads to widespread reflection upon the fundamental questions of economic activity: what does 'wealth' mean? How to define human wellbeing? How to apply the existing market logic to the global commons? How to combine climate resilience, economic growth and sustainability in development strategies in the vulnerable regions? There is increasing recognition that climate change is no longer so much a scientific issue or a knowledge-deficit problem as it is a social, cultural and ideological dilemma. It requires societies to redefine their identities, adjust their ambitions to the environmental realities and re-model their behaviours. In the case of the Western countries, the environmental crisis is just one of the potentially devastating contemporary problems, along with social apathy, growing discontent and increasing fragmentation. Although there is no clarity about the detailed configuration of the links within this complex system of phenomena, all its major elements—the predominance of growth-oriented development based on fossil fuels, neoliberal capitalism and the model of representative democracy—are being questioned. It seems that in the West common perception of the dramatic costs connected with civilizational and material development has already become a salient feature of late-modern culture.

Bibliography

Akerlof, G., & Kranton R. (2000) "Economics and Identity." *Quarterly Journal of Economics*, 115(3): 715–753.

Arendt, H. (1951) *The Origins of Totalitarianism*. New York: Schocken Books.

Arvidsson, A., & Peitersen, N. (2013) *The Ethical Economy: Rebuilding Value After the Crisis*. New York, NY: Columbia University Press.

Bäckstrand, K., & Kronsell, A. (Eds.) (2015) *Rethinking the Green State: Towards Climate and Sustainability Transitions*. Abingdon: Routledge.

Beck, U. (1999) *What is Globalization?* Cambridge: Polity Press.

Bickerstaff, K., & Walker, G.P. (2002) "Risk, Responsibility, and Blame: An Analysis of Vocabularies of Motive in Air-Pollution(ing) Discourses." *Environment and Planning A*, 34: 2175–2192.

Bina, O. (2013) "The Green Economy and Sustainable Development: An Uneasy Balance?" *Environment and Planning C: Government and Policy*, 31(6): 1023–1047.

Beck, U. (1992) *The Risk Society: Towards a New Modernity*. London: Sage.

Blauwhof, F.B. (2012) "Overcoming Accumulation: Is a Capitalist Steady-State Economy Possible?" *Ecological Economics*, 84: 254–261.

Brown, L. (2006) *Plan B 2.0: Rescuing a Planet Under Stress and a Civilization in Trouble*. Washington, DC: Earth Policy Institute.

Burke, M.B., Miguel, E., Satyanath, S., Dykema, J.A., & Lobell, D.B. (2009) "Warming Increases the Risk of Civil War in Africa." *PNAS*, 106(49): 20670–20674.

Clegg, S. (1989) *Frameworks of Power*. London: Sage.

D'Alisa, G., Demaria, F., & Kallis, G. (Eds.). (2014) *Degrowth: A Vocabulary for a New Era*. London: Routledge.

Dobson, A. (2003) *Citizenship and the Environment*. London: Routledge.

Eckhardt, G.M., & Bardhi, F. (2015) "The Sharing Economy Isn't About Sharing at All." *Harvard Business Review*. Available at: https://hbr.org/2015/01/the-sharing-economy-isnt-about-sharing-at-all (Accessed 20.12.2019).

Ferguson, P. (2015) "The Green Economy Agenda: Business as Usual or Transformational Discourse?" *Environmental Politics*, 24(1): 17–37.

FitzRoy, F., Franz-Vasdeki, J., & Papyrakis, E. (2012) "Climate Change Policy and Subjective Well-Being." *Environmental Policy and Governance*, 22(3): 205–216.

Frey, B.S. (2008) *Happiness: A Revolution in Economics*. Cambridge: MIT Press.

Frey, B.S. (2018) *Economics of Happiness*. Cham: Springer.

Friedman, T. (2008) *Hot, Flat and Crowded: Why the World Needs a Green Revolution—And How We Can Renew Our Global Future*. New York: Farrar, Straus & Giroux.

Foucault, M. (1980) *Power/Knowledge: Selected Interviews and Other Writings*. New York: Pantheon.

Gandini, A. (2016) *The Reputation Economy: Understanding Knowledge Work in Digital Society*. Basingstoke, UK: Palgrave Macmillan.

Gerber, E.W., & Hui, J. (2013) "Crowdfunding: Motivations and Deterrents for Participation." *ACM Transactions on Computer-Human Interaction* (TOCHI), 20(6): 1–32.

Giddens, A. (2006) *Nowoczesność i Tożsamość. „Ja" i Społeczeństwo w Epoce Późnej Nowoczesności*. Warszawa: PWN.

Gore, A., Jr. (1992) *Earth in the Balance*. Boston: Little Brown.

Green Economy Coalition (2012) *The Principles of a Green Economy: Initial Results of a Global Consultation*. London: Green Economy Coalition.

Guevarra, L. (2010) 'Gen Y's Green Demand for the Workplace'. Retrieved May 14 2019 from: http://www.greenbiz.com/news/2010/05/19/gen-ygreendemandsworkplace?utmsource=feedburner&utm_medium=feed&utm_campaign=Feed%3A+Greenbuzz+%28GreenBiz+Feed%29.

Hardin, G. (1968) The Tragedy of the Commons. *Science*, 162(3859): 1243–1248.

Hardin, G. (1998) Extensions of "The Tragedy of the Commons". *Science*, 280(5364): 682–683.

Harms, M. (2007) "What Drives Motivation to Participate Financially in a Crowdfunding Community?" Available at: http://www.grin.com/en/ebook/181293/what-drivesmotivation-to-participate-financially-in-a-crowdfunding-community (Accessed 21.12.2019).

Harris, P.G. (2010) *World Ethics and Climate Change: From International to Global Justice*. Edinburgh, UK: Edinburgh University Press.

Hemer, J. (2011) A Snapshot on Crowdfunding: Working Papers Firms and Region, No. R2/2011. Available at: http://www.econstor.eu/handle/10419/52302.

Hulme, M. (2010) "Cosmopolitan Climates Hybridity, Foresight and Meaning." *Theory, Culture & Society*, 27(2–3): 267–276.

Jackson, T. (2009) *Prosperity Without Growth? The Transition to a Sustainable Economy*. London: Sustainable Development Commission.

Lane, R. (1978) "Autonomy, Felicity, Futility." *Journal of Politics*, 40(Winter): 1–24.

Lindblom, Ch. (1977) *Politics and Markets*. New York: Basic Books.

Linnerooth-Bayer, J., Mechler, R., & Pflug, G. (2005) Refocusing Disaster Aid. *Science*, 309(5737), 1044–1046.

Mabey, N. (2007) *Delivering Climate Security: International Security Responses to a Climate Changed World*. London: Royal United Services Institute and Routledge. Whitehall Paper 69.

McKinsey Company (2018) "True Gen: Generation Z and Its Implications for Companies." Available at: https://www.mckinsey.com/industries/consumer-packaged-goods/our-insights/true-gen-generation-z-and-its-implications-for-companies (Accessed 28.11.2019).

McLaren, A., & Agyeman, J. (2015) *Sharing Cities: A Case for Truly Smart and Sustainable Cities.* Cambridge, MA: MIT Press.

Mobjörk, M., Gustafsson, M.-T., Sonnsjö, H., van Baalen, S., Dellmuth, L.M., & Bremberg, N. (2016) *Climate-Related Security Risks: Towards an Integrated Approach*. SIPRI.

Nanda, N. (2009) Climate change and trade policy: A preliminary discussion. *Asia Pacific Research and Training Network on Trade (UN ESCAP)*, (4).

Nordhaus, W.D. (2018) "Climate Change: The Ultimate Challenge for Economics, Nobel Lecture in Economic Sciences." Available at: https://www.youtube.com/watch?v=h1RkSuAs03Q (Accessed 22.11.2019).

Norgaard, K.M. (2011) *Living in Denial: Climate Change, Emotions, and Everyday Life*. Cambridge: The MIT Press.

Norton, B. (2002) "The Ignorance Argument: What Must We Know to Be Fair to the Future?" in D. Bromley and J. Paavola, eds., *Economics, Ethics, and Environmental Policy: Contested Choices*. Oxford: Blackwell: 35–52.

OECD (2011) Towards Green Growth.

OECD (2015) Income Inequality: The Gap Between Rich and Poor.
Ojala, M. (2005) "Adolescents' Worries About Environmental Risks: Subjective Wellbeing, Values, and Existential Dimensions." *Journal of Youth Studies*, 8(3): 331–347.
Ostrom, E. (2009) A Polycentric Approach for Coping with Climate Change. Policy Research Working Chapter 5095. World Bank.
Parry, M.L., et al. (2007) Climate Change 2007: Impacts, Adaptation and Vulnerability. Contribution of Working Group II to the Fourth Assessment Report of the Intergovernmental Panel on Climate Change. Cambridge: Cambridge University Press.
Pollin, R. (2019) "Advancing a Viable Global Climate Stabilization Project: Degrowth Versus the Green New Deal." *Review of Radical Political Economics*, 51(2): 311–319.
PricewaterhouseCoopers (2015) *Consumer Intelligence Series: The Sharing Economy*. Available at: https://www.pwc.com/us/en/technology/publications/assets/pwc-consumer-intelligence-series-the-sharing-economy.pdf (Accessed 19.11.2019).
Roberts, J.T., & Parks, B.C. (2007) *A Climate of Injustice: Global Inequality. North–South Politics, and Climate Policy*. Boston: MIT Press.
Ryan, R.M., & Deci, E.L. (2000) "Self-Determination Theory and the Facilitation of Intrinsic Motivation, Social Development, and Well-Being." *American Psychologist*, 55(1): 68–78.
Sage, G. (2007) "Sport and Social Resistance," in G. Ritzer, ed., *Blackwell Encyclopedia of Sociology*. Blackwell, Oxford: 4707–4710.
Schumpeter, J.A. *Capitalism Socialism and Democracy*. New York: Harper, 1942.
Sen, A. (1977) "Rational Fools: A Critique of the Behavioral Foundations of Economic Theory." *Philosophy and Public Affairs*, 6: 317–344.
The Nielsen Company (2015) *The Sustainability Imperative*. Retrieved April 20 2019 from http://www.nielsen.com/us/en/insights/reports/2015/thesustainability-imperative.html.
van Staveren, I. (2001) *The Values of Economics: An Aristotelian Perspective*. London: Routledge.
Stern, N. (2007) *The Economics of Climate Change*. Cambridge: Cambridge University Press.
Taeihagh, A. (2017) "Crowdsourcing, Sharing Economies and Development." *Journal of Developing Societies*, 33(2): 191–222.
Taylor, P., & Keeter, S. (2010) *Millennials: A Portrait of Generation Next*. Washington, DC: Pew Research Center.

United Nations Framework Convention on Climate Change (2012) Draft Report of the Conference of the Parties (COP) on Its Eighteenth Session. Doha, Qatar: UNFCCC.

United Nations Human Development Programme (2007) Human Development Report 2007–2008. Basingstoke: Palgrave Macmillan.

Waters, M. (1995) *Globalization*. London: Routledge.

World Bank (2009) *World Development Report 2010: Development and Climate Change*. World Bank.

Vanderheiden, S. (2008) *Atmospheric Justice: A Political Theory of Climate Change*. New York: Oxford University Press.

Veblen T.B. (1909) "The Limitations of Marginal Utility." *Journal of Political Economy* 17: 620–636.

8

Sustainable Energy Policies and Programs in Yakutia

Daria Gritsenko

Introduction

The circumpolar North is rich in energy resources. According to the much-cited US Geological Survey from 2009, the Arctic holds 13% of the world's undiscovered conventional oil and 30% of undiscovered conventional natural gas resources. Yet, many Arctic communities depend on imported diesel, experience energy poverty, and environmental vulnerability (Poelzer et al. 2016). This problem is very acute in Russia where many communities are isolated from the national or regional grid and rely on expensive diesel generation to cover their energy needs (Berdin et al. 2017).

Given the high energy cost, the negative environmental and health effects, and the increased availability of technology that can be used in the severe Arctic weather conditions, sustainable energy can be an answer to the local energy challenges (Cherniak et al. 2015; Poelzer et al. 2016). Sustainable energy for the purpose of this chapter means a combination

D. Gritsenko (✉)
University of Helsinki, Helsinki, Finland

E. V. Shabliy et al. (eds.), *Discourses on Sustainability*,
https://doi.org/10.1007/978-3-030-53121-8_8

of energy efficiency (EE)—reducing wasteful consumption—and uptake of renewable energy (RE) sources. In the past decade, pilot projects for switching remote villages from diesel-generated to wind- and solar-diesel hybrid power were realized in Canada, Russia, Greenland, and the US (arcticrenewableenergy.org). These projects demonstrated that together, EE and RE can reduce the demand for and supply of energy generated from fossil fuels (such as natural gas, oil, and coal-fired power plants). Yet, we still observe significant disparities in the scale and speed of Arctic and sub-Arctic sustainable energy transition (Mortensen et al. 2017).

Since 2013, the Russian federal government adopted new policy instruments in support of renewables, aiming to improve domestic energy security and, according to the official strategies, mitigate climate change (Smeets 2017). Yet, despite its huge natural potential and the existence of policy instruments, Russia derives less than 1% of its energy from renewable sources. According to some scholars, Russia is slow to adopt RE because it cannot be used as an "energy weapon" to gain leverage in international power struggles (Koch and Tynkkynen 2019). Others suggest that Russia's backwardness in RE uptake is due to the lack of appropriate policy instrumentation (Boute 2016). Currently, financial mechanisms for RE support exist only for the capacity market (since 2013) and in retail (since 2015), while no mechanism exists to support microgeneration and individuals cannot sell to and buy from the grid. What appears particularly interesting are the differences in the scale of RE and EE technologies uptake between the Russian regions (Gritsenko 2017). Among the regions that have made relative progress on EE/RE is the Republic of Sakha (Yakutia), the largest subnational unit in the Russian Federation located in Eastern Siberia and largely in the Arctic zone.

This chapter explores the perspectives of regional policy-makers, private companies, and civil society in order to understand the current status and near-future prospects of renewable energy policy and development of energy efficiency in Yakutia. It thereby contributes to the emerging body of research on the subnational tendencies in Russian energy policies and governance (Boute 2016; Salonen 2019). Using data from policy documents (laws, strategies, programs, official governmental websites) and a series of qualitative expert interviews, it provides an insight into the current and future challenges and opportunities with regard to developing renewable energy policy and programs in Yakutia. Twenty semi-structured interviews have been conducted in August 2018 with stakeholders who have been involved in EE and RE projects development in Yakutia. All interviews have been recorded and later transcribed in full. Thematic analysis, a method for identifying, analyzing, and reporting themes within data (Clarke et al. 2015), has been used to analyze the interview data. Reviewing the responsibilities, activities, and images for renewable energy in Yakutia, the chapter also provides a broader outlook at the role of the regions in shaping conditions that allow them to take advantage of the environmental, social, and economic benefits of incorporating renewable energy into their energy mix.

The chapter proceeds as follows. Section "Background on Yakutia" provides a general background on Yakutia. Section "Energy Generation and Provision in Yakutia" outlines the current system of energy generation and provision, in particular its economic and logistical aspects. Section "Mechanisms for Sustainable Energy Support in Yakutia" outlines legislative, financial, and institutional mechanisms that support sustainable energy in Yakutia. Section "Challenges to Sustainable Energy Development in Yakutia" analysis challenges for RE and EE policies and programs, while Sect. "Concluding Remarks" concludes.

Background on Yakutia

The Republic of Sakha (Yakutia) is located in Eastern Siberia, Russia. Yakutia is one of Russia's national republics, meaning that in accordance with the Russian Constitution it has its own legislature (called *Il Tumen*), Constitution, and is governed by the elected *Head of the Republic* (before 2014 called *President*). The republic is divided in uluses (rayon) and naslegs (municipalities)—administrative units rooted in traditional land and clan divisions. The regional capital—Yakutsk—is 4885 km away from Moscow.

Table 8.1 presents some basic economic characteristics of Yakutia. The republic occupies more than three million sq. km and thereby constitutes the largest subnational unit of the Russian Federation and the largest

Table 8.1 Characteristics of Sakha Republic (Yakutia) in 2015

Region	Area. sq. km	Population (1.1.16), thousands	GRP[a] in current prices, rubles	GRP[a] per capita in current prices, rubles	Extractive industries, % of GRP[a]	Rail road density, km/km^2	Road density, km/km^2	Ginni
Sakha Republic (Yakutia)	3083.5	960	749987.00	782629.40 (12,902 USD)	48.2	2	3.8	0.398

Source Author based on Rosstat—Russian Federation Federal State Statistics Service (www.gks.ru)
[a]Gross regional product (GRP) in current prices in rubles, average annual exchange rate in 2015: 1 USD = 60.66 RUB

indigenous territory in the world. The population of Yakutia is 960,000 (1.1.2016), composed of several ethnic groups, including Yakut (60%), Russians (30%), and Indigenous peoples of the North, such as Evenks, Evens, Dolgans, Yukaghir, and Chukchi. Most of Yakutian territory is located in extreme weather conditions, and on permafrost. Five, and soon 13 uluses are included into the Arctic Zone of the Russian Federation. The average population density is three persons per km^2, and in the Northern areas less than one person per km^2.

Residents of the remote Arctic communities face a number of challenges, including a lack of transportation infrastructure (road density 3.8 km/km^2, railroad density 2 km/km^2). The Republic, especially the Arctic uluses, has a strong reliance on imports from the South using sealift (large-scale transportation by sea), inland waterways or aircraft. Almost half of the regional GDP (GRP) comes from extractive activities (48% in 2015): the Republic is engaged in diamond, gold, silver, coal mining, as well as small scale gas extraction. As a result, Yakutia can be characterized as a peripheral, extractive, sparsely populated region (Table 8.1). The Republic receives almost 50% of its budget through federal transfers.

The climatic and geographical characteristics of the Republic significantly complicate the functioning of the Yakutian energy system (Northern Forum, 2017). First, the climatic conditions are extreme, with a large seasonal difference of temperatures from +35 °C in summer to −50 °C in winter. Second, the service area exceeds 3 million sq. km, with uneven population density and industrial consumers clustered around point source resources (so-called focal industrial development), so that private and industrial consumption can seldom be coupled spatially. Third, the absence of year-round roads in most of the republic's territory leads to seasonality of fuel, material, and technical supply. As a result, a unique energy system has emerged in Yakutia.

Energy Generation and Provision in Yakutia

Centralized and Decentralized Power Supply

Yakutia is divided into centralized and decentralized power supply zones (Programma 2017).[1] The centralized energy supply zone consists of three energy districts—Western, Central, and Southern. 99.6% of energy in the Western region comes from hydro energy, the remaining 0.4% from natural gas. In the Southern district, coal makes up 99.3% of consumption, with 0.7% remaining for diesel. In the Central district, 97% of demand is covered by the natural gas, and 3% by diesel fuel. As a result, power generation in Yakutia relies on a diverse energy mix. These energy districts developed independently for a long time, and only in 2019, South and West have been connected to the Russian grid via Irkutsk, providing interoperability with the nation-wide grid and allowing the Republic to join the wholesale market. The centralized power supply covers 36% of the territory of the republic, where 85% of the population lives.

Decentralized energy supply is characteristic of the *Isolated systems in the North*. Yakutia has 144 power plants isolated from each other or any other grid that supply individual villages and mining enterprises. The main part of the capacity of autonomous power plants (over 200 MW) is located in the Northern Energy District. The autonomous power industry covers an area of 2.2 million km^2 (64%) and mainly relies on diesel-electric stations. Yet, as of 01.01.2019, 13 out of 144 isolated settlements had supplementary renewable energy sources installed. About 15% of the Yakutian population (roughly 144000 people) live within the isolated energy area.

The total installed capacity of power plants in the republic is 2853.3 MW, mainly relying on thermal (42% of the total capacity) and hydroelectric (34% of the total capacity) power plants (Table 8.2). Diesel power plants in the structure of installed capacity occupy 24%. Currently, renewables occupy 0.05% of generation—plan is to have 4% by 2022.

In terms of actors, electricity provision in Yakutia is dominated by two state-owned enterprises: "Yakutskenergo," a subsidiary of the federal

Table 8.2 Power generation in Yakutia by type in 2014 and 2015

Power generation type, mW	Year	
	2014	2015
Hydro	957.5	957.5
Thermal	1246.2	1201.2
Diesel	657.4	693.3
Wind	0.25	0.04
Solar	0.225	1.34
Total installed capacity, mW	2861.6	2853.3

Source Author based on Programma (2017)

monopolist company—JSC "Energy Systems of Russia," that is responsible for three energy regions, and its subsidiary "Sakhaenergo" that maintains diesel generators in isolated areas (Russian JSC Energy Systems of the East, 2016). Table 8.3 shows that three generating companies (Yakutskenergo, Sakhaenergo, DGK—selling to Yakutskenergo in the South, all daughters of Rushydro) supply 69% of the total capacity.

Table 8.3 Power generation in Yakutia by actor (in 2015)

N	Company	2015	%
1	Yakutskenergo (ПАО «Якутскэнерго»)	1226.2	0.429748
2	DGK (АО «ДГК»)	618	0.216591
3	Vilyuy HES-3 (АО «Вилюйская ГЭС-3»)	277.5	0.097256
4	Sakhaenergo (АО «Сахаэнерго»)	188.2	0.065959
5	Surgutneftegaz (ОАО «Сургутнефтегаз»*)	152.7	0.053517
6	Transneft (ПАО «Транснефть»)	92	0.032243
7	Alrosa (АК «АЛРОСА» (ПАО))	50.7	0.017769
8	Almazy Anabara (АО «Алмазы Анабара»)	33.9	0.011881
9	Burenergo (ООО «Бурэнерго»)	25.8	0.009042
10	Taas-Yuryakh Neftegazdobycha (ООО «Таас-Юрях Нефтегаздобыча»)	20	0.007009
11	Gazprom Burenie (ООО «ГАЗПРОМ БУРЕНИЕ»)	23	0.008061
12	YaTEK (ОАО «ЯТЭК»)	15	0.005257
13	Elgaugol (ООО «Эльгауголь»)	11.5	0.00403
14	Seligdar (ПАО «Селигдар»)	4.7	0.001647
	Others	114.1	0.039989
	Total installed capacity	2853.3	100

Source Author based on Programma (2017)

The only private supplier HES-3 belongs to Alrosa—and is generating for Alrosa (Vilyuiskaya GES, n.d.). All the remaining power generation companies indicated in Table 8.3 (NN5-14) are resource companies that built their own generation power to lower the cost of electricity generation.

When it comes to heat generation in Russia, three-quarters of it is centralized and delivered through the district heating system that has been installed in the Soviet Union. While there are various models, including municipal unitary enterprise (MUP), a state unitary enterprise (GUP), and cooperatives, regulation of heat energy generation mainly takes place at the municipal level. Noticeable exceptions are private houses that can be heated individually with wood or boilers (gas, oil, pellets). Because heat generation systems are out of date, they are inefficient and prone to regular emergencies. In remote areas of the Arctic and the Far East, including the Yakutian Northern Energy District, the heat component reaches up to 65% of the total cost of housing and communal services, becoming its most expensive part (Salonen 2019).

Heat generation is dominated by the State Unitary Enterprise "Housing and Communal Services of the Republic of Sakha (Yakutia)" (GUP ZHKH), established by decree of the President of the Republic in 2003 by joining the housing and communal services enterprises operating in the regions of the Republic. The creation of a cenralised enterprise was motivated by frequent accidents on municipal enterprises. GUP ZHKH supplies thermal energy to ¾ of Yakutia, enjoying an almost monopoly besides a few exceptions (MUPs and in isolated districts, Sakhaenergo does heating, too). Yet, as several Interviewees informed, around 40% of Yakutian households use wood stoves and individual boilers, and recently electric heating was also gaining traction as electricity grid has been expanded to more uluses.

Energy Tariffs

Energy economics in Russia is directly connected to energy generation and provision. Most of the Russian North-West, Centre, South are connected into a nation-wide grid maintained by RAO UES—United

Energy Systems (Boute 2012). In these territories pricing is competitive, wholesale-based auction. Some parts are isolated from this nation-wide grid and rely on their own electricity generation. In these parts, pricing is based on tariffs—regulated prices established by state authorities. At the federal level, the Federal Tariff Service regulates wholesale tariffs, while Regional Energy Commissions regulate retail prices. The Russian government sets a cap on electricity prices for the regions annually, and the regional regulator sets the tariffs for the residents and companies within these limits. Until January 2019, Yakutia was not connected to the nation-wide grid, but had its own power systems and completely belonged to the regions with tariff regulation. Since the main bulk of materials for this study has been collected prior to 2019, and the gird connection has only been realized for one year at the moment of writing this chapter, it appears difficult to assess its influence on the RE/EE dynamics, and hence for the purpose of this investigation, Yakutia will be treated as a region with tariff regulation.

Up until 2017, in the Republic of Sakha (Yakutia), electric power industry was financed through four types of cross-subsidies: between electric and thermal energy, between categories of consumers, between four energy regions of the republic (Central, Western, South Yakutia, and Northern), and between the Republic of Sakha (Yakutia) and ECO East (Programma 2017). Unified tariffs were set for electricity generation regardless of the source. In 2016, the "average" tariff after subsidization was 6.15 RUR/KWh. At the same time, the economically feasible electricity tariff in the Northern grid area was 36.31 RUR/KWh and, in the central grid zone, it was 4.34 RUR/KWh. The burden of covering the difference between the feasible and the set tariff mainly fell on industrial consumers, covering costly cross-subsidization of diesel energy (in 2016 amounted to total of 6.8 billion rubles) and also gradually choosing other options (building own energy generation capacity to get out of the system).

In 2017, a new tariff equalization between the Price Zones and Tariff Zone has been introduced at the federal level. In essence, the three-year program of equalizing energy tariffs to the average Russian level meant subsidizing Far Eastern energy from the wholesale electricity market. The leveling mechanism was aimed at all industrial consumers of Chukotka,

Magadan region, Kamchatka, Sakhalin, and Yakutia, without any division into categories. Already after one year, tariff equalization has had a significant effect on savings both for business and for regional budgets. Yet, the expected economic effects (increase of industrial consumption, new industries) have been questioned. In late 2019, the Russian federal government decided to keep the program unchanged for 2020, as well as to continue tariff equalization until 2028 but in a modified form, that will be decided upon in 2020 (eastrussia.ru 25.11.2019).

Energy Logistics

Energy logistics for the centralized power supply zones of Yakutia mainly rely on local fuel delivery schemes, as has been clarified in the research interviews. The majority of coal for the Southern Energy District is delivered by rail from the section "Neryungri" by the supplier of JSC holding company "Yakutugol" The Central Energy District relies on transit supply of natural gas, with major suppliers including OAO Yakutgazprom, OAO Lenagaz, and CJSC ALROSA-Gaz. In the Western Energy District, supply is maintained by local hydroelectric production at the Vilyuy power plant.

The fuel logistics in the isolated parts of the Republic is considerably more complex than in the centralized energy zone, as follows from the subsequent information derived from the research interviews. Almost the entire volume of diesel fuel is delivered from outside the region. First, the delivery from refineries are conducted by rail to a number of transhipment hubs: Tommot petroleum depot (for early delivery of diesel fuel along the Aldan River), Station "Neryungri – passenger" (for consumers of the Southern Energy District), Ust-Kut petroleum depot (the main transshipment base for further shipment of petroleum products by river navigation). Also the sea routes are in use. Murmansk and Arkhangelsk ports in Western Arctic are key transhipment hubs for diesel fuel delivery from Northern sea route to Tiksi, as well as fuel storages along the Yana, Indigirka, and Kolyma rivers. Second, from the transhipment hubs, further delivery is carried out to diesel power plants or tank farms of JSC Sakhaneftegazsbyt, a regional government agent for the import and storage of petroleum products. Finally, winter roads ("zimniki") are used for transportation of petroleum products from points of accumulation (usually along the rivers) to the final consumers in remote settlements. The volume of transportation of petroleum products by the winter roads is ca. 25–30 thousand tons annually. This fuel delivery scheme is expensive due to the high distances from fuel suppliers, seasonality of fuel delivery, high number of transhipments, high fuel storage cost, and extremely large service area. As a result, cost optimisation of energy provision in the isolated parts of Yakutia is a persistent economic problem and an issue on the republican political agenda.

Mechanisms for Sustainable Energy Support in Yakutia

Legislative Mechanisms

Yakutia has closely followed the Russian federal legislation concerning energy efficiency and renewables. In 2009, the Russian federal Energy Efficiency Law N261 ("On Energy Conservation and Improving Energy Efficiency, and Introducing Amendments to Certain Legislative Acts of the Russian Federation") has been adopted, and only a year later in Yakutia, a Republican law "On measures for the implementation of energy conservation and energy efficiency in RS (Ya)" has been implemented. The federal law focused on providing incentives for broader uptake of energy-efficient technologies to improve the state of the Russian economy, competitiveness, and energy security. Among the main instruments, it introduced mandatory EE requirements, tracking of EE in state and municipal procurement, tax incentives for energy conservation initiatives, and administrative liability for non-compliance. In addition, the Energy Efficiency Law established an obligation to adopt regional and municipal programs, to which Yakutia complied. The Yakutian Energy Efficiency Programme and Roadmap set two key goals to be achieved by 2020: to decrease the energy intensity of the gross regional production by 40% and to decrease the consumption of energy resources by 2.7 mln tones (of fuel) equivalent by 2020, with the baseline indicators taken for 2007. While the final results of the program have not yet been published, it has been assessed that between 2011 and 2016, it allowed saving almost 760 mln Rub of the regional energy budget (Ammosova 2016).

In 2014, Yakutia has become the first Russian sub-national unit to adopt a law on Renewable Energy, following the Russian federal law adopted in 2013. The Yakutian RE law states improvement of energy security, reliability, savings, as well as environmental protection as its goals. The document has a framework character, listing a wide range of measures that could be used by the Republican government and the local administrations to create favorable conditions for energy efficiency and renewables, such as state programs and targets, tax benefits, special tariffs. Yet, there is no concrete implementation plan and the law has so far been not successful to attract outside funding for renewable projects.

In 2015, the Yakutian government launched POLE—a republican Program for Optimisation of Local Energy 2016–2025—that consisted of a detailed analysis of all isolated settlements in the Republic (144) and considered how to modernize their energy systems. The POLE programme (2015) set three key objectives: (1) Providing high-quality and reliable power supply to isolated settlements; (2) Reducing the amount of subsidies to local energy; and (3) Upgrading fixed assets in unsatisfactory technical condition. For each generating facility, various reconstruction options are considered, such as replacement and/or modernization of diesel generators, installation of renewable energy sources (wind and solar stations), or development of the electric grid complex. As a result, five scenarios for Yakutia's isolated energy systems modernization are formulated in POLE. The first two scenarios are based on the modernization of diesel generators, the third scenario prioritized the development of electricity grid, the fourth scenario is focused on the development of renewables, and the fifth combines grid connectivity and renewables (Table 8.4). The Yakutian government has favored the most comprehensive (#5) scenario, yet, it did not possess the financial resources for its implementation. The RF Ministry of Energy (Minenergo), whom the Yakutian government asked for financial assistance, was only ready to back the cheapest alternative (#2). Eventually, the Ministry

Table 8.4 Five scenarios for Yakutia's isolated energy systems modernization (POLE)

Scenario number	1	2	3	4	5
CAPEX (bln rub, 2016 prices)	13.06	11.96	29.29	20.8	36.98
Decrease in cross subsidy (bln rub, 2016 prices)	21.8	22.5	36.6	25	28.9
Decrease in economically feasible tariff	13.9%	14.4%	16.0%	17.3%	18.8%
Power generation efficiency	34%	34%	34%	39.2%	29%
Decrease in fuel delivery	44%	44%	50%	58%	63%

Source Author based on POLE Programme (2015)

refused to finance POLE, arguing that Yakutia is not the only region with isolated systems (though, the largest) and the federal state prefers a complex country-wide solution rather than support of one specific region. In 2017, such a solution has been adopted in the form of tariff equalization, while POLE was put on ice.

Financial Mechanisms

The main support mechanism for renewable energy in Russia—power supply contracts to the wholesale market at a guaranteed price—was introduced in 2013. This mechanism is meant for the capacity market and primarily aims at conducting a competitive selection of investment projects for the construction of generating facilities operating on the basis of renewable energy and the closing of power supply contracts for selected projects. This mechanism is not applicable to the isolated systems. In 2015, the Russian Federal Government adopted support scheme for renewable energy in retail electricity markets, including isolated zones that are not connected to the grid. This mechanism creates an obligation for grid companies to buy energy at regulated tariffs. Yet, the development of renewable energy in the retail market is limited by tariff uncertainty. Since the tariff is set upon commissioning the power plant, it is difficult for the potential investors to accurately assess their return on investment. As already noticed, there is currently no financial mechanisms in Russia that would support microgeneration, for instance, by private individuals. Moreover, Russian regions do not have the capacity to set up their own financial support mechanisms. As a result, Yakutia, among other regions, currently does not have any republican programs that would financially stimulate the development of renewables.

The situation is somewhat better in the area of energy efficiency. In 2007, the federal *Housing and Utilities Reform Facilitation Fund* was established as a state corporation mandated to provide financial support to the regions and municipalities for the overhaul of apartment buildings and the relocation of citizens from emergency housing stock. Since 2013, the Fund has also been granted responsibility for financing the upgrading of utility infrastructure systems. Since that time, it regularly runs special campaigns to support energy effective management of housing stock and the introduction of resource-saving technologies. In the field of thermal energy generation, saving, and management, the Fund supports projects that integrate renewable energy solutions.

On the private side, energy efficiency projects can be realized through energy service companies (ESCO). The Russian ESCO market started to develop after the federal energy efficiency legislation has been enacted in 2009. Usually, energy service contract is done for 3–5 years, the ESCO company brings capital required for the energy-saving measures, completes the necessary work, and returns the investment and generates profit. This is based on the premise that the tariff for the budget organization that goes under ESCO is fixed for the whole period of the contract. Once ESCO is finalized, the organization is left with a more efficient energy system and lower bills. Concession and Payment by installment are another two instruments that can be used to finance energy efficiency initiatives. These instruments have so far been less popular than ESCO. The energy service contract allows to modernize the equipment, the lighting system (heating), and pay off the savings, but as soon as we are talking about municipal property used to provide the population with heat (water) and other consumers, then the concession agreement mechanism is used. Concessions are larger, more comprehensive than ESCOs, offer tax breaks, while payments-by-installment are smaller than ESCO (typically 1–1.5 y payback time), used when small modifications or equipment are needed.

Institutional Mechanisms

To support the implementation of the energy efficiency programs, the Regional Agency of Energy Saving (RAER) has been created by the Yakutian government in 2010. RAER is an institution mandated to oversee the implementation of the regional energy efficiency program and is expected to provide comprehensive technical, financial, and knowledge resources for the implementation of energy efficiency measures in municipalities, the public sector, and the state-run utilities. The ultimate goal of RAER is to help the state budget of Yakutia to decrease its energy-related expenditures for utilities through the implementation of comprehensive energy-saving measures to optimize the fuel and energy balance. To reach this goal, the Agency performs the functions of the state contractor for energy saving. It is entitled to deliver paid services, such as conducting energy surveys, installing and maintaining metering devices, providing feasibility studies, and developing energy-saving programs. In addition, the Agency has a for-profit energy service company that can take on any of the projects developed by the Agency.

While there is no comparable institution that would provide support to renewable energy development, in 2012–2018 an annual International conference "Renewable Energy Development in the Far East of Russia" has been held in Yakutsk as a cooperation between Rushydro and the Yakutian government. According to the Conference website, the goal of the organizers was to show that renewable energy can be economically efficient in isolated villages. The Conference instantly became a platform for capacity building as it allowed local authorities, engineering specialists, businesses, and other stakeholders to meet, establish business contacts, exchange new knowledge and relevant information on available support tools, technical solutions and initiatives, as well as to sign project and cooperation agreements. Moreover, the Conference was international and allowed investors and businesses from other countries to get acknowledged with the realities of the implementation of renewable energy projects in the Russian Arctic and Far East. The Conference also brought concrete results. In 2012, a framework agreement between the Yakutian government and "RAO ES of the East" led to the construction of the world's largest polar SES in Batagai.

In 2017, the republican government and the Hevel group of companies signed an Agreement on cooperation in the field of power supply to isolated settlements of the Sakha Republic (Yakutia) through the construction of autonomous hybrid power plants based on solar energy (Rusnano 2017). Same year, the Japanese government organization NEDO, together with the Government of the Kamchatka Territory and RAO ES of the East, signed an agreement that led to the construction of a unique wind power complex in Ust-Kamchatsk. The Conference, nevertheless, has been discontinued in 2018 and the reasons for that have not been announced.

Challenges to Sustainable Energy Development in Yakutia

As pointed out in the Introduction to this chapter, despite the humble results in an international comparison, Yakutia positively stands out in its sustainable energy initiatives among the Russian regions. In what follows, I analyse opportunities and challenges for sustainable energy in Yakutia as they have been identified by the experts I interviewed in August 2018.

The key challenges to expanding renewable energy policy and programs are high cost, lack of adequate technology, and lack of supporting institutions. These three challenges are intrinsically linked with each other. High cost of renewable energy stems largely from the lack of adequate technology: due to seasonality and imperfect battery technology, in the Arctic climate effective storage of energy produced from renewables is not economically feasible, and hence most of the existing facilities do not have a storage option. This, in turn, means that supplementary technology, usually a diesel generator, is still required to provide robust energy supply to the remote areas. The use of RE as a complement to diesel generation can enable savings, but the startup costs are generally too high to justify installations, in particular under the conditions of fossil fuel subsidies and lack of financial and institutional mechanisms that would explicitly support the uptake of RE. While the instrumentation and financial basis for the implementation of energy efficiency initiatives are fairly developed, there remain significant gaps in government responsibilities with regard to renewable energy, both at the federal and at the republican level. There is no accessible public funding for RE projects: neither the federal nor the republican governments have a dedicated RE fund. The existing projects rely on market-based instruments and are financed through private investments. Lack of credible future commitments with regard to progressive policy development in the area of renewable energy is limiting private investors' interest. An additional complexity is brought in by the fact that Russia is reliant on imported technology, but following the financial crisis and the international sanctions, some technology providers became unavailable.

There is also a number of disincentives that stall the development of sustainable energy in Yakutia. First, Russia as a producer country maintains relatively low price for hydrocarbons. In today's conditions, the economics of RE has to be perfect to play out. While the situation can change, most experts agree that the current situation does not incentivize the investors to think about the more distant future. In particular, the system of energy tariff equalization adopted in 2017 has made the short- and mid-term investment outlook for renewables unattractive. Even though the Yakutian government emphasizes the importance of a long-term perspective upon the republican energy system, which is reflected in their preference toward the comprehensive POLE scenario, low fuel prices and absence of available funding create a situation of path-dependency.

Second, the state in Russia traditionally acts as a guarantor of heat and light (Tynkkynen 2019), and it grants significant energy subsidies to regional and local authorities. As a result, many parties have no interest in sustainable energy, in particular, the resource providing organizations whose budgets are fully subsidized. This creates a complex relationship between the regional authorities who are interested in energy savings and resource organizations who see energy savings as a threat to their existence. Also municipalities have no responsibility, nor markets, nor resources to engage with sustainable energy: while wasteful energy practices can be covered through subsidies, energy savings do not bring real savings to the budgets, but only remove the subsidies which cannot be repurposed for other developmental needs. The absence of drive for resource efficiency and self-sufficiency at the local level is deeply ingrained into the system of local government in Russia. The subsidy regime thereby subverts the common sense, making key organizations into losers rather than the winner of the sustainable energy transition.

Finally, the topics of community, culture, and education were raised by the interviewees as critical points in the development of sustainable energy. First, under the current legislation, it is difficult for individual consumers to realise financial benefits from investing into the renewables as individuals remain consumers, not prosumers—meaning, they cannot sell their production surplice to the grid, and the only way they can benefit is by saving on their energy expenses. For people residing in apartment housing, also the saving option cannot be realized. Given the relatively high cost of installations, the existing subsidy regime, and extensive gasification program realized across Russia by Gazprom (Tynkkynen 2016), the economic interest for an individual to invest in sustainable energy is limited. As communities are seldom involved in local decision-making, there is also a lack of awareness with regard to the social and environmental benefits of sustainable energy generation. In order for communities to be willing to take responsibility for their energy production and consumption, knowledge inequalities need to be alleviated. The interviews provided several examples of how the realization of sustainable energy projects in local schools and kindergartens has had a positive effect not only on children who thereby learned about renewable energy and sustainability, but also on their parents. There is anecdotal evidence that in these localities households demonstrate higher interest and willingness in installing renewables.

Concluding Remarks

This chapter provided a brief outlook regarding the current status and near-future prospects of renewable energy policy and the development of energy efficiency in Yakutia. It has shown that the state authorities currently play a key role in framing the playing field for sustainable energy development. At the federal level, the state provides the overall legislative framework, which then can become a starting point for the regional policy initiatives. These, however, are significantly limited due to the highly centralized energy governance and the marginal role of municipal administration in Russia (Kropp et al. 2017). The case of tariff equalization is a case in point that demonstrates how the centralized energy systems disadvantage local governments and leads to little financial incentive to get involved in RE projects for potential private investors.

Yet, we should not disregard the important role the Yakutian government played in the development of both EE and RE projects in the Republic. First, it had an enabling legislative function since the regional by-laws were essential to the implementation of ESCO in Yakutia. Second, it paid attention to attracting international investors, ensuring the exchange of best practices, sharing of knowledge and information between the state, business, and scientific sectors by providing communication platforms, in particular, an international conference. Finally, in some cases, the non-interference of the regional government has been crucial to the success of small municipal and community initiatives. At the same time, the regional government experienced a number of challenges, in particular relating to the accessibility of funding.

As a consequence, the progress in the area of sustainable energy that one can witness in Yakutia has mainly been achieved through the engagement of private business based on republican mechanisms (mainly, ESCO) or research and development (R&D) grants. ESCO model has worked well for projects aimed at increasing energy efficiency (upgrade of municipal buildings through insulation, automation, installation of solar collectors), while R&D model has mainly been applied to launch electricity generation with the help of solar or wind power stations, aiming at testing the potential for fuel saving in isolated communities. Table 8.5 shows that each model has its own benefits and shortcomings, as well as its own effects on the local power dynamics.

Though Yakutia has been percieved as a pioneer of sustainable energy in the Russian Arctic, since 2018, further developments retarded. Hence, the success of the sustainable energy policies and programs depends on the opportunities to integrate them within the existing energy governance frameworks. So far, the attempts at such integration—as expressed

Table 8.5 Benefits and shortcomings of the two key models for sustainable energy promotion

	Benefits	Shortcomings	Power dynamics
R&D	Relatively flexible form of support, no need to bring ROI, freedom to experiment	Not scalable, one off projects	Tends to benefit the incumbents, potentially stimulates corruption
ESCO	Purely commercial projects, both ESCO provider and receiver benefit financially	Savings (additional resources) benefit the contractor, end user often left aside	Alleviates local resource constraints, but does not empower the local community

Source Author's compilation based on research interviews

in the POLE programme and the Yakutian Energy Strategy 2018–2022—have not brought any significant results and no comprehensive policy framework supported by financial resources exists. The challenges Yakutia faces are not unique for this region: all across Russia, regional sustainable energy policies and programs are highly dependent on the federal level of government. Yet, the regional mechanisms that have been installed in Yakutia continue to bring up change, even if at a very limited scale. In sum, the regional governments' attempts to make their energy systems more resilient by improving infrastructure, reducing cost, and seeking energy efficiency, are conditioned, but not determined, by the center-region power relations.

Note

1. All the data presented in the Section has been derived from the State program of the Republic of Sakha (Yakutia) "Development of the energy sector of the Republic of Sakha (Yakutia) for 2018–2022" (https://glava.sakha.gov.ru/ot-27-noyabrya-2017-g-----2231) and Sakhastat, the Territorial body of the Federal State Statistics Service for the Republic of Sakha (Yakutia) (https://sakha.gks.ru/).

Bibliography

Ammosova, V. (2016). Energosberegayushchie tekhnologii s'ekonomili v Yakutii 570 millionov rublei (in Russian). https://www.yakutia.kp.ru/online/news/2615043/.

Berdin, V. Kh., Kokorin, A. O., Yulkin, G. M., & Yulkin, M. A. (2017). *Renewable Energy in Off-Grid Settlements in the Russian Arctic*. Moscow: WWF.

Boute, A. (2012). Promoting Renewable Energy through Capacity Markets: An Analysis of the Russian Support Scheme. *Energy Policy*, *46*, 68–77.

Boute, A. (2016). Off-Grid Renewable Energy in Remote Arctic Areas: An Analysis of the Russian Far East. *Renewable and Sustainable Energy Reviews*, *59*, 1029–1037.

Cherniak, D., Dufresne, V., Keyte, L., Mallett, A., & Schott, S. (2015). *Report on the State of Alternative Energy in the Arctic*. Ottawa: Carleton University.

Clarke, V., Braun, V., & Hayfield, N. (2015). Thematic Analysis. *Qualitative Psychology: A Practical Guide to Research Methods*, 222–248.

eastrussia.ru (25.11.2019). Выравнивание энерготарифов для ДФО сохранят на 2020 год. https://www.eastrussia.ru/news/vyravnivanie-energotarifov-dlya-dfo-sokhranyat-na-2020-god/.

Gritsenko, D. (2017). A Local Perspective on Energy Development in the High North. *Aleksanteri Insight-Snapshots of Eurasia*, *2017*(1). https://www.helsinki.fi/en/news/sustainability/a-local-perspective-on-energy-development-in-the-high-north.

Koch, N., & Tynkkynen, V. P. (2019). The Geopolitics of Renewables in Kazakhstan and Russia. *Geopolitics*, 1–20 (online first).

Kropp, S., Aasland, A., Berg-Nordlie, M., Holm-Hansen, J., & Schuhmann, J. (Eds.). (2017). *Governance in Russian Regions: A Policy Comparison*. Springer.

Mortensen, L., Hansen, A. M., & Shestakov, A. (2017). How Three Key Factors Are Driving and Challenging Implementation of Renewable Energy Systems in Remote Arctic Communities. *Polar Geography*, *40*(3), 163–185.

Northern Forum. (2017). Interview with Gavril Prokopievich Levin, Minister of Housing, Public Utilities and Energy of Sakha Republic (Yakutia). https://www.northernforum.org/en/news/298-the-government-of-sakha-republic-yakutia-considering-high-rates-of-development-and-introduction-of-renewable-energy-sources-in-yakutia-believes-that-further-work-in-this-area-is-useful-indeed.

Poelzer, G., Hoogensen Gjorv, G., Holdmann, G., Johnson, N., Magnússon, B. M., Sokka, L., & Tsyiachniouk, M. (2016). Developing Renewable Energy in Arctic and Sub-Arctic Regions and Communities [Online]. Retrieved September 26, 2019, from https://research.wur.nl/en/publications/developing-renewable-energy-in-arctic-and-sub-arctic-regions-and-.

POLE Programme. (2015). https://docplayer.ru/52711906-Programma-optimizacii-lokalnoy-energetiki-respubliki-saha-yakutiya.html.

Programma. (2017). The State Program of the Republic of Sakha (Yakutia) "Development of the Energy Sector of the Republic of Sakha (Yakutia) for 2018–2022." https://glava.sakha.gov.ru/ot-27-noyabrya-2017-g—-2231.

Rusnano. (2017). http://www.rusnano.com/about/press-centre/news/20170623-yakutiya-i-khevel-zaklyuchili-soglasheniye-o-stroitelstve-solnechnykh-elektrostantsiy.

Russian JSC Energy Systems of the East. (2016). *RAO Energy Systems of the East. Developing energy for development of Far East*. http://www.rao-esv.ru/en/activity/geography/.

Salonen, H. (2018). Public Justification Analysis of Russian Renewable Energy Strategies. *Polar Geography*, *41*(2), 75–86.

Salonen, H. (2019). Modernization of Russian District Heating Systems with the Help of Biomass Energy—A Gordian Knot? Environmental Innovation and Societal Transitions. https://doi.org/10.1016/j.eist.2019.10.006.

Smeets, N. (2017). Similar Goals, Divergent Motives: The Enabling and Constraining Factors of Russia's Capacity-based Renewable Energy Support Scheme. *Energy Policy*, *101*, 138–149.

Tynkkynen, V. P. (2016). Energy as Power—Gazprom, Gas Infrastructure, and Geo-governmentality in Putin's Russia. *Slavic Review*, *75*(2), 374–395.

Tynkkynen, V. P. (2019). *The Energy of Russia: Hydrocarbon Culture and Climate Change*. Cheltenham, UK and Northampton, MA, USA: Edward Elgar.

Vilyuiskaya GES. (n.d.). http://www.shpp.ru/activity/sales_activity.aspx.

9

Diaspora and Renewable Energy in Manubhai Autobiography *Tide of Fortune*

Shilpa Daithota Bhat

The idea of being environment-friendly goes beyond economic and tax incentives. The desire to be venerable to nature, to strike a balance between technology and verdure, are themes that persist not only in the context of science and economics but also in its representation in written discourses which makes the domain a critical framework for the study. This edited anthology explores the several angularities of the subject of sustainability and renewable energy that exist in various disciplines and my chapter seeks to investigate this aspect through the study of the autobiography *Tide of Fortune* (2008) by Manubhai Madhvani. What is significant about this narrative is that the writer displays sensitivity towards environmental protection through intelligent ways for renewing factory wastes. An autobiography quintessentially revolves around the author's life. However, Manubhai Madhvani takes pride in arranging and creating best practices for renewing factory remains and looking for ways to creatively manufacture energy. While, the subject of renewable sources of energy is seen as falling under the category of technology, through

S. D. Bhat (✉)
Ahmedabad University, Ahmedabad, India

E. V. Shabliy et al. (eds.), *Discourses on Sustainability*,
https://doi.org/10.1007/978-3-030-53121-8_9

Madhvani's narrative, what is highlighted is new ways of perceiving energy and sensitizing society on the theme, through the process of writing.

The subject of ecology requires maximum attention and creating this awareness has to happen in a variety of ways. While scientists and innovators help design approaches to resolving the challenges of discovering renewable sources of energy; disseminating such information in society becomes a major need, to be able to implement such technology. To this end, journalists and writers can help and contribute meaningfully. This chapter, therefore, seeks to bring to centre stage, the idea of narratives as critical in disseminating and sensitizing society towards the necessity of conserving and creating energy.

Manubhai Madhvani was an entrepreneur whose ancestral homeland was India and whose ancestors went to Africa to establish business. Over a period of time, their family business took off and they were able to open new branches of factories not only in Africa but also in the rest of the world, owing to the forced expulsion by the Ugandan dictator, Idi Amin.[1]

It is through this trajectory of international diasporic movements and the spirit of business and entrepreneurship that I discuss the representation and discussion of renewable energy and ecological sensitivity, in the narrative under examination. It is necessary to mention here that while written narratives are frequently studied for the value they present in terms of retrieving history, politics, and sociology, their contribution with regards to representation of energy resources has been hardly undertaken. As works read by individuals and hence an avenue to trigger sensitization of people towards their environment, it is crucial that this dimension of representation of energy resources and their possible renewal is taken account of. It is this gap that the essay predominantly endeavours to bridge and contextually, in the objective of interdisciplinary exploration of the terrain of renewable energy. Interestingly, Madhvani suggests this idea of renewable resources through the disposal of bagasse, which is waste remain after producing sugar. He proudly discusses this in the work and underlines that there is no wastage of resources and everything is converted into energy. Therefore, there is this cycle of using and reusing substances, so as to conserve and reproduce energy that not only saves money but also ensures protection of the environment, preventing degradation of ecology.

Entrepreneurship and Thematic Background for Renewable Energy in the *Tide of Fortune*

The essential idea of the argument in this essay is that, an autobiography that generally dominantly concentrates on the life and experiences of the author, includes the technical processes and methods of renewing industrial waste in the factories owned. To a reader, this might seem anachronistic. Nonetheless, these gestures at the attentiveness of the author and the industrialist are not only profit-making but also the effect that his factories ultimately have on the environment. These have far-reaching consequences on human health and ecology. Naturally, the emphasis on renewable sources of energy, go beyond the costs saved in the process, since the stakes in terms of health and environment are significantly high and long term in nature.

In the *Tide of Fortune* (2008), Madhvani describes the migratory routes, goals, and intentions of the Gujarati community in Africa. It was a discovery for the Gujaratis to find a geographical region that provided fertile land and ample resources for the farming and production of sugar. Gunvantrai Acharya in his seminal fictional representation of the migratory journey of the Gujaratis to Africa in *Dariyalal* (2000), depicts the sea-faring merchants in the seventeenth century, who travelled to and fro, from Gujarat state in India to Africa. The western part of India was located strategically near the sea and people found it easy to consider the idea of travelling to Africa in huge 'dhows', that is, huge boats with lateen sails. Of course, these journeys were beset with life-threatening dangers and risks. Nevertheless, the undying spirit to operate business and profit, motivated the Gujarati community to explore the then unknown land of Africa. It is also necessary to note that the Gujaratis had no political aspirations for territorial control, unlike the other European forces who 'scrambled for Africa'. The Asian Indian Gujarat did not have any interest in tearing up the continent for the resources. Their only objective was the engage in trade in a resourceful manner. It is also noteworthy that the Gujaratis were strongly connected to their cultural roots and did not give up their language, culture, cuisine, and so on. Gijsbert Oonk's (2013) in his research study used the terms 'settled strangers' to highlight that the Gujarati community was settled in terms of carving out sources of living but culturally they did not deeply hybridize with the host society. Hence, they were 'settled' yet remained 'strangers' in Africa. Chhaya Goswami in her *Globalization Before Its Time: Gujarati Traders in the Indian Ocean* (2016) describes in detail the successful overseas business and trade (particularly in ivory, spices, pearls, etc.) carried out by the Gujaratis in Africa. This history is critical in understanding the relations between the people of both the countries and the European colonizers whose presence in Africa, complicated processes of political control.

Madhvani underscores the fascinating journey of the Gujarati Indian diaspora that migrated to different parts of the world and engaged in trade, entrepreneurship, and business. The suggestion is that the Gujarati community in East Africa had a penchant for pioneering efforts in starting new ventures and technologies in their businesses. The autobiography under examination portrays the migration and settlement of the Madhvani family in East Africa and the trials and tribulations they went through for their survival and success. They strongly established themselves in Uganda. The other popular business enterprise was run by the Nanji Kalidas Mehta family. *Tide of Fortune* is a fascinating autobiography of the Madhvani family and is also a work that powerfully depicts the history, politics, and economic structure in the East Africa of the twentieth and twenty-first centuries. Muljibhai Madhvani—Manubhai's father travelled to East Africa in 1904, when the region was under the domain of the colonizing British force (and was hence called British East Africa). In the beginning, Muljbhai opened a retail shop at Jinja, just about 54 miles from Uganda's capital, Kampala. From this point, the family soon purchased about 800 acres of land to establish sugar plantation farms and sugar factory. Their journey from Gujarat to Uganda is a fascinating saga characterized by the spirit of taking risks and quick decisions, tempered with a desire to follow ancestral traditions and culture. The spirit of entrepreneurship ran in the family and Manubhai

Madhvani was quick to take advantage of the knowledge and experience of his forefathers, as also the opportunity to gain formal education. While Manubhai successfully ventured into his family business, what is noteworthy is his penchant for renewing energy out of the waste produced from sugar and glass factories. This thoughtfulness towards environmental protection is critical in understanding the role of business ventures in shaping awareness in society. Therefore, the representation of such themes in written work is significant since the information circulated can further intensify the notion of possibilities of renewable energy and the direction that organizations can take up to step up efforts in this potential domain. Verhoef (2017) states that 'The resilience of the Madhvani Group is an example of entrepreneurial survival, which manifested in a number of African economies after Africa turned to a more democratic political order and the opportunities of the market towards the end of the twentieth century'.[2]

Diaspora, Indentured Labour System, and Colonialism

In his depiction of the history of sugar, Manubhai Madhvani gestures at the infamous indentured labour system evolved by the colonizing British force as a reinvention of the earlier African institution of slavery. After the Slavery Abolition Act of 1833, the British recast the system, through the establishing of the indentured or contract labour system in which the colonized populations of the world were shifted to the sugar plantation farms in British colonies, to serve as labour under contract. The colonized population became a source of cheap labour. This system resulted in large scale migration of 'native' population from their homelands to new and alien hostlands. While Manubhai suggests that there was association of slave labour with sugarcane plantation farms, the indentures labour system which was notorious for its violence against the workers is not referred to anywhere, except that there is just a cursory mention to the historical and political episode. Perhaps, this has something to do with the idea that the Gujaratis and the Madhvani family, in particular, had nothing to do with the exploitative tactics pursued by the British colonial forces. When the Gujaratis were setting up flourishing business enterprises, the Britishers (and other European colonial forces) were extending their colonial control over Africa. The Gujaratis at this point of time were often seen as 'Asian Shylocks' and prototype imperialists—occupying a space that was rather 'in-between' the European colonizer and the native Africans. Interestingly, the Gujaratis were able to negotiate between both the races and those who were born and brought up in Africa felt more African rather Asians. Gijsbert Oonk (2013) in his research study on the subject of Gujarati migration uses the term 'global citizens' to describe the phenomenon of Gujaratis who previously travelled from Gujarat to Africa and from here dispersed to the rest of the world. Manubhai states that:

> In the fifteenth century, European sugar was refined in Venice, though it remained a small-scale business as it was difficult to transport. Then in 1492, the Genoese explorer Christopher Columbus sailed to America and took sugar cane seedlings to plant in the Caribbean. The climate there was so advantageous to the growth of cane that a large industry was quickly established. In the late seventeenth century, sugar production became dependent on and interdependent with the slave industry, as plantation owners began to rely on slave labour.
>
> By 1750, there were 120 sugar refineries in Britain, producing 30,000 tonnes per annum. So vast were the profits that sugar was called 'white gold' and the government taxed it highly. Only in 1874 was this tax abolished and sugar prices came within the means of ordinary citizens.[3]

This bit of history is necessary to understand how sugar producers negotiated between politics and economics and understood them; and how they dealt with the challenges of ecology and business within their organizational setup. It is also noteworthy that Manubhai Madhvani worked in tandem with the locals in Uganda. Nevertheless, the term indenture or similar terms suggesting the likes of colonial contract labour has never been historically used in the context of workers in the sugarcane plantations farms run by Gujarati entrepreneurs. However, Manubhai traces the history of sugar production in his memoir and states that 'Cane Sugar was probably first used in Polynesia, from whence it spread to India.

In 510 BC, Emperor Darius of Persia invaded India, where he found "the reed which gives honey without bees". That is to say, sugar cane, a very tall grass with thick stems...grown in tropical countries...looks like bamboo or elephant grass...up to five metres tall...grows in eighteen months'.[4] In outlining the history of sugar invention and production, and the politics that allowed for the dissemination of the technology, Manubhai also happens to chart the migratory routes of the sugar distribution. The desire to make profits, made Emperor Darius to keep the sugar production a secret. Nonetheless, when the Arab invaders entered Persia in AD 642, they discovered the long-secret and learnt how to produce the crop and farmed sugar in North Africa and Spain. It was the Crusades of the eleventh century that resulted in the Europeans knowing about the production of sugar and the product reached England in 1099.[5] Manubhai states that 'The first sugar in England was recorded in 1099 and it was very much a luxury. For instance, in AD 1319, sugar was available in London at "two shillings a pound", which would be about US$ 100 a kilo today'.[6] The Madhvani family was swift in gauging the significance of entering the business in Uganda and created conditions that were favourable to the growth of sugar.

Renewable Energy in Sugar Production in the *Tide of Fortune*

What is notable in Manubhai's work is the underscoring of the idea of how the production of sugar could result in zero ecological destruction and in fact, resourcefulness in terms of production and conservation of energy. The author achieves this through interweaving industrial and technical processes within his autobiographical narrative. While describing the steps involved in the sugar production method, he claims that 'The interesting thing about producing sugar is that very little is wasted. The bagasse is burned...the mol can be fermented and distilled to make ethanol, and the soil that covers the cut cane is returned to the fields'.[7] Manubhai shows the sensitivity and excitement of being able to avoid environmental degradation through careful disposal of bagasse. El-Newehy, Barghini, Cometa, Miertus, and Chiellini (2012) state that

'The bagasse derived as a residue from this activity is usually burned to produce the electricity required by the sugar mills and, in some cases, excess energy is even sold in the electricity market.... However, a great amount of bagasse is still wasted'.[8] Baikow (1992) in his study entitled *Manufacture and Refining of Raw Cane Sugar* mentions that when the fiber content of sugar cane is very high or there is a shortage of boilers and oil must,

> ...be burned, large quantities of bagasse may accumulate. The handling of unused bagasse with bulldozers and its removal by trucks and spreading on fields is quite expensive. A more practical procedure in this case is to burn the bagasse as soon as there is excess...an inefficient and sometimes dangerous procedure. Alternatively, the excess of bagasse can be discharged from the bagasse conveyor into a hopper or bin and then blown out through a duct with a powerful blower. At the exit from the duct the bagasse can be ignited and kept burning as long as bagasse is being blown out.[9]

Manubhai highlights that the bagasse is 'burned in large boilers to generate steam and electricity for the factory: high pressure steam is used to drive electricity-producing turbines, which power the rotating mills. The steam then becomes low pressure steam for cleaning sugar crystals.... The steam produced by burning bagasse is the sole source of energy in our sugar factory. Steam drives the turbines, which either turn the mills or produce electricity'.[10] Mahler (1981) in his research study looks at 'the relevance of dependency hypotheses to a concrete historical relationship'.[11] He states that 'sugar has figured very prominently in North-South relations since the dawn-of the modern era, and close analysis of the sugar trade should offer a good idea of how one mode of North-South relations has evolved over centuries. Moreover, Britain, the first capitalist nation, was an important participant from the earliest days in the expansion of European influence into what is now called the Third World'.[12] The connection between the sugar economy, colonization, competition between the contending European forces in the 'Third World' underlines the interest that prevailed in the sector. It was during this historical phase that the Gujaratis had migrated to East Africa and established themselves in the sugar production business. They negotiated between the colonial forces and the native Africans to further their business interests. Unlike the European countries, the Gujaratis did not have political interests. Their aim was never political expansion and control. They merely wanted to make profits out of their business ventures. This they tried to achieve through interactions with the European imperialists as well as the Africans. It is necessary to observe that this is a very distinct difference between the Asian Gujaratis and the colonizers. Within the business ventures, the Gujaratis tried their best to incorporate the latest technologies and expertise not only in terms of sugar plantation and production but also in terms of finding methods to convert the disposable bagasse into energy production and thereby conservation. This aspect is critical in understanding that this mode of energy production helped save expenditure on harnessing or gaining energy. At the same time, it took care of the environment.

Glass Production and Renewable Energy[13]

Manubhai Madhvani diversified his business and his family engaged in a variety of product specializations in different parts of the world. Whichever domain, the family chose for its business expansion, they made sure that they not only focused on the use of modern technology but also the sensible use of discarded remains through conversion to energy that both saved costs on the procuring of energy and avoided contamination and destruction of the environment.

> Just as the steam in a sugar factory is recycled, so the broken or malformed glass inadvertently produced in a glass factory is put to good ecological use. Rejected bottles from the production process are carried on conveyors to the cullet crusher (cullet from outside can also be collected and used in the process, thus contribution to a cleaner environment), from whence they are transported to a storage silo in the batch-mixing plant. They are then fed into the batch, as described above. So, anything, rejected gets recycled.[14]

Manubhai Madhvani suggests consistency in his interest in ecological sensitivity and renewing energy through disposable waste from his factories. This attitude is an important cue in the contemporary world, at a time when environmental degradation is at its peak. One of the greatest challenges that the world is grappling with today is reuse of wastes like plastic that is disposed everywhere, including oceans, endangering aquatic life.

Conclusion

Manubhai Madhvani's autobiography is important in understanding the drive required to not only engage in business with profit motives, but also to look at the successful recycling of unwanted residues. The fact that Madhvani represents this information in his autobiography that prominently focuses on his family's journey from Gujarat to Africa, yet considers it necessary to talk about renewable sources of energy and possibilities of such processes, is intriguing as well as critical. Autobiographies predominantly discuss an individuals' life and experiences. However, Madhvani captures the technological processes that he perceives as important in narrating his life in the book.

Notes

1. The rule of Idi Amin in Uganda brought untold sufferings and struggle in the life of Asians. Those who had already purchased immovable property lost their wealth. Money saved in the banks were confiscated. There was nothing left for the Asians to survive. Amin's policy of 'ethnic cleansing' impelled Asians to travel to Britain first since Africa was 'British Africa' and therefore Asians were treated as British subjects. Eventually, however, Britain changed its policy and Asians from Africa had to look for new migratory destinations and they went to Canada, the US, and Australia.
2. Grietjie Verhoef, *The History of Business in Africa: Complex Discontinuity to Emerging Markets* (Cham, Switzerland: Springer, 2017), 107.
3. Manubhai Madhvani and Giles Foden, *Tide of Fortune: A Family Tale* (Noida: Random House, India, 2009), 78.

4. Ibid., 77.
5. Ibid., 77–78.
6. Ibid., 78.
7. Manubhai, *Tide of Fortune: A Family Tale*, 84.
8. El-Newehy Mohamed, Arianna Barghini, Stefania Cometa, Stanislav Miertus, and Emo Chiellini "Polymeric Materials from Renewable Resources: Blends of Poly (3-Hydroxybutyrate) and Cellulose Acetate Derived from Rice Straw and Bagasse." In *Renewable Resources and Renewable Energy: A Global Challenge* (Boca Raton: CRC Press, 2012), 161.
9. V. E. Baikow, *Manufacture and Refining of Raw Cane Sugar* (Amsterdam: Elsevier, 1992), 82.
10. Manubhai, *Tide of Fortune: A Family Tale*, 83, 85.
11. Vincent A. Mahler, "Britain, the European Community, and the Developing Commonwealth: Dependence, Interdependence, and the Political Economy of Sugar." *International Organization* 35, no. 3 (1981): 470.
12. Ibid.
13. Manubhai Madhvani provides a fascinating discussion on the history of glass prior to his discussion on the reusing of waste remains after the process of production. He states "It was probably first discovered naturally, the fusion having been caused by the impact of a meteorite, or of lightning, or from the massive heat of volcanic lava. Obsidian, a type of naturally occurring glass, was used to create sharp-bladed knives by the ancient Egyptians, the Aztecs, and others. However, glass was first made in Iraq or on the Syro-Palestinian coast. Glassmaking was a Semitic speciality….The Roman scholar, Pliny the Elder (23–79), tells the story of the discovery of glass at the sandy mouth of the Belus River, near present-day Haifa. The tradition is that a merchant ship laden with nitrum being moored at this place, the merchants were preparing their meal on the beach, and not having stones to prop up their pots, they used lumps of nitrum from the ship, which fused and mixed with the sands of the shore, and there flowed streams of a new translucent material, and thus was the origin of glass." Please see Manubhai. *Tide of Fortune: A Family Tale*, 192.
14. Manubhai, *Tide of Fortune: A Family Tale*, 195.

Bibliography

Acharya, Gunvantrai P., and Kamal Sanyal. 2000. *Dariyalal*. Calcutta: Dictum in association with Thema.

Baikow, V. E. 1992. *Manufacture and Refining of Raw Cane Sugar*. Amsterdam: Elsevier.

Clay, Rebecca. "Renewable Energy: Empowering the Developing World." *Environmental Health Perspectives* 110, no. 1 (2002): A30–33. http://www.jstor.org/stable/3455287.

Collins, Lyndhurst. "Renewable Energy from Energy Crops: A Global Perspective." *Geography* 84, no. 2 (1999): 169–79. http://www.jstor.org/stable/40573272.

El-Newehy Mohamed, Arianna Barghini, Stefania Cometa, Stanislav Miertus, and Emo Chiellini. "Polymeric Materials from Renewable Resources: Blends of Poly (3-Hydroxybutyrate) and Cellulose Acetate Derived from Rice Straw and Bagasse." In *Renewable Resources and Renewable Energy: A Global Challenge*. Boca Raton: CRC Press, 2012.

Fornasiero, Paolo, and Mauro Graziani. *Renewable Resources and Renewable Energy: a Global Challenge*. Boca Raton [etc.]: CRC Press/Taylor & Francis Group, 159–188.

Goldemberg, José. "Ethanol for a Sustainable Energy Future." *Science* 315, no. 5813 (2007): 808–10. http://www.jstor.org/stable/20038951.

Goswami, Chhaya, and Jaithirth Rao. 2016. *Globalization Before Its Time: The Gujarati Merchants from Kachchh*. London: Penguin Books.

Kerr, Richard A. "Do We Have the Energy for the Next Transition?" *Science*, New Series, 329, no. 5993 (2010): 780–81. http://www.jstor.org/stable/40799726.

Luiz Pinguelli Rosa, and Suzana Kahn Ribeiro. "Avoiding Emissions of Carbon Dioxide through the Use of Fuels Derived from Sugar Cane." *Ambio* 27, no. 6 (1998): 465–70. http://www.jstor.org/stable/4314770.

Machado, Pedro. "A Forgotten Corner of the Indian Ocean: Gujarati Merchants, Portuguese India and the Mozambique Slave-Trade, c.1730–1830." *Slavery & Abolition* 24, no. 2 (2003): 17–32.

Madhvani, Manubhai, and Giles Foden. 2009. *Tide of Fortune: A Family Tale*. Noida: Random House India.

Mahler, Vincent A. "Britain, the European Community, and the Developing Commonwealth: Dependence, Interdependence, and the Political Economy of Sugar." *International Organization* 35, no. 3 (1981): 467–92. http://www.jstor.org/stable/2706432.

Mehta, Nanji Kalidas.1987. *Dreams Half-expressed: An Autobiography*. Mumbai: Vakil and Sons.

Morrison, Angela R. "The New Federal Renewable Fuel Standard Program Rules and Potential Obstacles for Implementation." *Natural Resources & Environment* 25, no. 4 (2011): 33–37. http://www.jstor.org/stable/23054850.

Oonk, Gijsbert. 2013. *Settled Strangers: Asian Business Elites in East Africa (1800–2000)*. New Delhi: Sage.

Pimentel, David, G. Rodrigues, T. Wang, R. Abrams, K. Goldberg, H. Staecker, E. Ma, L. Brueckner, L. Trovato, C. Chow, U. Govindarajulu, and S. Boerke. "Renewable Energy: Economic and Environmental Issues." *BioScience* 44, no. 8 (1994): 536–47. https://doi.org/10.2307/1312281.

Sharma, Monika. 2014. *Socio-Cultural Life of Merchants in Mughal Gujarat*. India: Partridge Pub.

Vassanji, M. G. 1996. *The Book of Secrets: A Novel*. New York: Picador USA.

———. 1999. *Amriika*. Toronto: McClelland & Stewart.

———. 2004. *And Home was Kariakoo: A Memoir of East Africa*. Toronto: Doubleday Canada.

———. 2004. *The In-between World of Vikram Lall*. New York: A.A. Knopf.

———. 2007. *The Assassin's Song*. New York: Alfred A. Knopf.

———. 2013. *The Magic of Saida*. New York: Random House.

Verhoef, G. 2017. *The History of Business in Africa: Complex Discontinuity to Emerging Markets*. Cham, Switzerland: Springer.

Visram, M G. 1990. *Allidina Visram: The Trailblazer*. Mombasa, Kenya: MG Visram.

Conclusion

Human-induced climate change is a pressing issue; and global energy security is one of the most important questions involving intercultural responsibility. It is crucial to critically look at these and other climate-related problems from a political, business, and scientific points of view and discuss the impact of economics and developing policies on climate change mitigation (efforts to reduce future climate changes) and adaptation (efforts to reduce the vulnerability of society to climate change impacts).[1] Climate change mitigation requires immediate action, and the serious problem related to fossil fuel use requires a prompt response; there is an urgent need to stop using fossil fuels completely. Climate adaptation and mitigation are interrelated, and adaptation efforts may be costly and less likely to succeed without mitigation initiatives.[2] In the 2014 Climate Change Impacts in the US report, mitigation is defined as

> Technological change and substitutions that reduce resource inputs and emissions per unit of output. Although several social, economic, and technological actions would reduce emissions, with respect to climate change,

E. V. Shabliy et al. (eds.), *Discourses on Sustainability*,
https://doi.org/10.1007/978-3-030-53121-8

> mitigation means implementing actions to reduce greenhouse gas emissions or increase the amount of carbon dioxide absorbed and stored by natural and man-made carbon sinks.

Various consulting companies, including the McKinsey consulting group, regularly issue reports on climate change.[3] Business as Usual Plan is no longer adequate and acceptable in the current scenario; climate change skepticism and denial violate human rights and youth rights. There are many questions associated with climate ethics; one of the related problems is whether or not future technological development could facilitate removing GHGs from the atmosphere, another complicated issue is related to a global tax on carbon emissions. The world has become interdependent, "so has the world of energy – the energy industry, its resources and shortages, its prices, and its pollution."[4] According to the World Bank Group climate change action plan,

> Current weather extremes already affect millions of people, putting food and water security at risk, and threatening agricultural supply chains and many coastalcities. Without further action to reduce extreme poverty, provide access to basic services, and strengthen resilience, climate impacts could push an additional 100 million people into poverty by 2030.[5]

GHGs, CO_2, and other gases, "cause Earth to be out of energy balance, that is the main driving force of human-induced climate change."[6] As James Jansen argues, GHGs cut must begin promptly, "or climate change will be pushed beyond a point at which changes proceed out of human control."[7] It is necessary to keep the average temperature increase below 2 °C, "which in turn requires rapid global action to reach zero net emissions by the second half of the century."[8] This volume is a discussion of climate change problems from multiple perspectives. Most chapters examine climate change mitigation and adaptation strategies and initiatives. Developed and developing countries are considered in this book that brings together leading researchers from around the world. The Paris Agreement established a goal on climate adaptation, which aims at strengthening national adaptation efforts, including international support and global cooperation. This adaptation perspective should develop into National Adaptation Plans and efforts.[9] Each climate plan reflects the country's ambition for reducing GHGs.[10]

Notes

1. Climate Change Impacts in the United States—Amazon S3. https://s3.amazonaws.com/nca2014/low/NCA3_Climate_Change_Impacts_in_the_United%20States_LowRes.pdf (last accessed 2/28/2020).
2. Ibid.
3. "Climate Risk and Response." https://www.mckinsey.com/~/media/McKinsey/Business%20Functions/Sustainability/Our%20Insights/Climate%20risk%20and%20response%20Physical%20hazards%20and%20socioeconomic%20impacts/MGI-Climate-risk-and-response-Full-report-vF.ashx (last accessed 2/23/2020).
4. Michael H. Dworkin and Benjamin K. Sovacool, *Global Energy Justice: Problems, Principles, and Practices* (Cambridge: Cambridge University Press), 29.
5. World Bank Group Climate Change Action Plan 2016–2020. https://openknowledge.worldbank.org/bitstream/handle/10986/24451/K8860.pdf?sequence=2&isAllowed=y, 1.
6. "Global Temperature in 2018 and Beyond," James Hansen et al. http://www.columbia.edu/~jeh1/mailings/2019/20190206_Temperature2018.pdf (last accessed 2/23/2020).
7. Ibid.
8. World Bank Group Climate Change Action Plan 2016–2020. https://openknowledge.worldbank.org/bitstream/handle/10986/24451/K8860.pdf?sequence=2&isAllowed=y, 7 (last accessed 2/23/2020).
9. "What Is the Paris Agreement?" https://unfccc.int/process-and-meetings/the-paris-agreement/what-is-the-paris-agreement (last accessed 2/23/2020).
10. "Nationally Determined Contributions (NDCs)." https://unfccc.int/process-and-meetings/the-paris-agreement/nationally-determined-contributions-ndcs#eq-1 (last accessed 2/23/2020).

Index

E. V. Shabliy et al. (eds.), *Discourses on Sustainability*,
https://doi.org/10.1007/978-3-030-53121-8

S

The manufacturer's authorised representative in the EU is Springer Nature Customer Service Centre GmbH, Europaplatz 3, 69115 Heidelberg, Germany. If you have any concerns regarding our products, please contact ProductSafety@springernature.com

Printed and bound by CPI Group (UK) Ltd, Croydon, CR0 4YY

30/09/2026

02225698-0010